STRUCTURAL DATA AND INVARIANTS OF NINE DIMENSIONAL REAL LIE ALGEBRAS WITH NONTRIVIAL LEVI DECOMPOSITION

R. CAMPOAMOR-STURSBERG

Nova Science Publishers, Inc.
New York

Library of Congress Cataloging-in-Publication Data
Campoamor-Stursberg, R.
Invariants of nine dimensional real Lie algebras with nontrivial Levi decomposition / R. Campoamor-Stursberg.
p. cm.
ISBN 978-1-60456-497-6 (hardcover)
1. Lie algebras. 2. Decomposition methods. 3. Invariants. I. Title.
QA252.3.C36 2008
512'.482–dc22 2008007303

Published by Nova Science Publishers, Inc. ✤ *New York*

Contents

Preface

This work grew up from the necessity of having reliable data on the generalized Casimir invariants of non-semisimple Lie algebras appearing in an ample spectrum of physical applications, like integrable systems, non-Abelian Yang-Mills theories or dynamical symmetry algebras. Although for semisimple algebras the solution to this problem is well known, for non-solvable and non-semisimple Lie algebras the analysis has been mainly focused on certain types of algebras. However, for some applications, like the classification of contractions, a systematic approach is necessary. For low dimensions this has been done by various authors, while for dimension nine, the last dimension to have been classified, almost no information is available.

The main objective is to analyze various structural properties of indecomposable nine dimensional real Lie algebras having a non-trivial Levi decomposition. These algebras, classified in the early 90s by P. Turkowski, actually show patterns that do not occur in Lie algebras of lower dimension. The aim of this analysis is to determine a fundamental system of invariants for the coadjoint representation, that is, the generalized Casimir invariants. Due to the large number of essential parameters, a direct analytical approach is not always useful, since there is an enormous number of special cases to be analyzed. In order to integrate the corresponding systems of PDEs, we consider the theory of semi-invariants developed by Trofimov and the reduction of the problem to certain subalgebras, the invariants of which can be computed easily or are already known. The procedure is loosely based on the labelling of representations using subgroup chains (the so called missing label operator problem), which allows to express the invariants of the total algebra as functions of invariants of subalgebras. The method also provides those Lie algebras that arise as an extension of some subalgebra, therefore providing information on the trivial co-homology.

Chapter 1

Introduction

Invariant functions of the coadjoint representation of Lie algebras constitute an important tool in representation theory and many applications, both mathematical and physical, such as completely integrable systems, differential equations or symmetry analysis in physical problems. In Representation theory, the eigenvalues of Casimir invariants are used to label irreducible representations of groups, as well as for the study of reduction chains with respect to subgroups. The semisimple case has been completely solved, mainly by the work of Casimir, Van der Waerden, Chevalley or Racah [18, 19, 38]. In this mathematical approach, polynomial invariants are proven to be elements in the centre of the universal enveloping algebra, and the number of such operators is given by the rank of the algebra. This classical result allows an extension to algebras having rational invariants, where these are obtained as ratios of semi-invariants of the coadjoint representation [19]. Extensive work has been done on the eigenvalues of Casimir operators of classical groups and their generating functions [4, 5, 34, 35]. For the case of non semisimple Lie algebras no general criteria for the number and structure of independent invariants exist, up to certain specific classes [1, 10, 14, 42]. Various methods have been developed to compute the invariants of Lie algebras. The classical study of the universal enveloping algebra is useful for the semisimple and reductive case, as well as some other classes of algebras [18, 48]. The Kirillov approach has been proven useful for certain types of semidirect products, although the most employed technique is the analytical method, i.e., the formulation of the problem by means of differential equations [21]. For specific classes of algebras some purely algebraic methods based on representation theory have been developed [13, 31], such as semidirect products of Heisenberg and semisimple algebras, the special affine algebras $\mathfrak{sa}(n, \mathbb{R})$ used in the metric-affine gravity theories or Borel subalgebras of simple Lie algebras [2, 14, 22, 31, 36, 42]. Recently a new method based on the moving frames of E. Cartan has been proposed [8]. This approach, which has proven to be of interest in the symmetry analysis of differential equations (see [28, 40] and references therein) has the advantage of avoiding completely the integration of differential equations, and uses the inner automorphism group of Lie algebras. This new algorithm provides considerable simplification in the description of invariants, and constitutes a powerful tool to analyze solvable Lie algebras in any dimension.

Invariants of Lie algebras have been computed for solvable Lie algebras up to dimension six ([8, 26, 29] and references therein) and in dimension 7 for nilpotent Lie algebras [39]. Some large classes of solvable algebras in any dimension have also been studied, such as triangular Lie algebras, algebras with abelian or Heisenberg radicals, or solvable algebras with graded nilradical of maximal nilpotence index [3, 41, 46]. For non-solvable Lie algebras various references on the structure and properties of invariants exist, and for important classes of algebras, such as the inhomogeneous and the special affine Lie algebras, the invariants have been computed explicitly [13, 14, 20, 31, 32, 36]. Many particular algebras of physical importance have also been considered, including the classical kinematical Lie algebras and their subalgebras [30] and algebras related to Cayley-Klein geometries [22].

In this work we determine a fundamental system of invariants for nine dimensional real indecomposable Lie algebras having a nontrivial Levi decomposition. The classification of these algebras was obtained in [45]. Although we use the classical analytical approach, direct integration of the invariants is a cumbersome task. In order to obtain the invariants, we consider the theory of semi-invariants for the coadjoint representation [42, 43] and the reduction to subalgebras the invariants of which can be computed easily or are already known. The procedure is based in the labelling of representations using subgroups chains (so called missing label operator problem) developed for arbitrary Lie algebras in [33].

Unless otherwise stated, any Lie algebra $\mathfrak{g}$ considered here is of finite dimension over the field $\mathbb{K} = \mathbb{R}$ and indecomposable, i.e., not splitable into a direct sum of ideals.

Chapter 2

Semidirect Sums of Lie Algebras

The classification of Lie algebras is reduced to specific classes due to the Levi decomposition theorem, which states that any Lie algebra is formed from a semisimple Lie algebra $\mathfrak{s}$ (called the Levi factor of $\mathfrak{g}$) and a maximal solvable ideal $\mathfrak{r}$, called the radical. It follows that the Levi factor $\mathfrak{s}$ acts on $\mathfrak{r}$, in either one of the two following forms:

$$[\mathfrak{s}, \mathfrak{r}] = 0 \tag{2.1}$$
$$[\mathfrak{s}, \mathfrak{r}] \neq 0$$

In the first case we obtain a decomposable algebra $\mathfrak{s} \oplus \mathfrak{r}$, whereas the second possibility implies the existence of a representation R of $\mathfrak{s}$ which describes the action, i.e., such that

$$[x, y] = R(x) \cdot y, \ \forall x \in \mathfrak{s}, y \in \mathfrak{r}. \tag{2.2}$$

We will use the notation $\overrightarrow{\oplus}_R$ to describe semidirect products. Since (2.2) implies that the radical is a module over $\mathfrak{s}$, we have to expect structural restrictions on the radical (for direct sums any solvable Lie algebra being suitable). Two important structural properties on semidirect products were obtained in [45]:

Proposition 1 *Let $\mathfrak{s} \overrightarrow{\oplus}_R \mathfrak{r}$ be a Levi decomposition of a Lie algebra $\mathfrak{g}$.*

1. *If R is irreducible, then the radical $\mathfrak{r}$ is abelian.*

> 2. *If the representation R does not possess a copy of the trivial representation, then the radical $\mathfrak{r}$ is a nilpotent Lie algebra.*

Indecomposable Lie algebras (i.e., that do not decompose as a direct sum of ideals) of dimension $n \leq 8$ and having non-trivial Levi decomposition were completely classified by Turkowski in [44], while the nine dimensional case was analyzed in [45]. All these algebras have a Levi part of rank one, isomorphic to the compact form $\mathfrak{so}(3)$ or the real normal form $\mathfrak{sl}(2,\mathbb{R})$.[1] In dimension nine, 63 isomorphism classes of indecomposable algebras, some depending on one or more parameters, were found. Additionally we define the Lie algebra $L_{9,7}^{*}$ defined by the nontrivial brackets

$$
\begin{array}{llll}
[X_1,X_2] = X_3, & [X_1,X_3] = -X_2, & [X_2,X_3] = X_1, & [X_1,X_4] = \tfrac{1}{2}X_7, \\
[X_1,X_5] = \tfrac{1}{2}X_5, & [X_1,X_6] = -\tfrac{1}{2}X_5, & [X_1,X_7] = -\tfrac{1}{2}X_4, & [X_2,X_4] = \tfrac{1}{2}X_5, \\
[X_2,X_5] = -\tfrac{1}{2}X_4, & [X_2,X_6] = \tfrac{1}{2}X_7, & [X_2,X_7] = -\tfrac{1}{2}X_6, & [X_3,X_4] = \tfrac{1}{2}X_6, \\
[X_3,X_5] = -\tfrac{1}{2}X_7, & [X_3,X_6] = -\tfrac{1}{2}X_4, & [X_3,X_7] = \tfrac{1}{2}X_5, & [X_4,X_6] = X_8, \\
[X_5,X_7] = X_8, & [X_4,X_9] = X_4, & [X_5,X_9] = X_5, & [X_6,X_9] = X_6, \\
[X_7,X_9] = X_7, & [X_8,X_9] = 2X_8. & &
\end{array}
$$

This algebra has the Levi decomposition

$$
L_{9,7}^{*} = \mathfrak{so}(3) \overrightarrow{\oplus}_{R_4 \oplus 2D_0} \mathfrak{g}_{6,82}^{2,0,0}, \tag{2.3}
$$

where the notation for the radical $\mathfrak{g}_{6,82}^{2,0,0}$ is that used in [24]. Since the real solvable Lie algebras $\mathfrak{g}_{6,82}^{2,0,0}$ and $\mathfrak{g}_{6,92}^{*}$ are non-isomorphic, the Lie algebras $L_{9,7}^{*}$ and $L_{9,7}^{p}$ of [45] are non-isomorphic for any p, showing that this algebra is missing in the list of [44].

---

[1] Algebras having a Levi part of higher rank are necessarily decomposable.

Chapter 3

Method for Computing Invariants

Among the possible approaches to compute the invariants of Lie algebras, we choose the analytical one. Let G be a connected Lie group, $\mathfrak{g}$ its corresponding Lie algebra and $Ad : G \to GL(\mathfrak{g})$ the adjoint representation. The coadjoint representation of G is given by the mapping:

$$Ad^* : G \to GL(\mathfrak{g}^*) : \left(Ad_g^* F\right)(x) = F\left(ad_{g^{-1}}x\right),$$

where $g \in G, F \in \mathfrak{g}^*, x \in \mathfrak{g}$. Denoting the space of analytical functions on $\mathfrak{g}^*$ by $C^\infty(\mathfrak{g}^*)$, we say that a function $F \in C^\infty(\mathfrak{g}^*)$ is a semi-invariant for the coadjoint representation if it verifies the condition

$$F\left(ad_{g^{-1}}x\right) = \chi(g)F(x)$$

for any $g \in G$, χ being a character of G. In particular, if $\chi(g) = 1$, then F is called an invariant of $\mathfrak{g}$. It follows at once that F is an invariant if and only if F is constant on each coadjoint orbit. The usual method to compute the (semi-)invariants of a Lie algebra is making use of the theory of linear partial differential equations [23]. Let $\{X_1, .., X_n\}$ be a basis of $\mathfrak{g}$ and let $\left\{C_{ij}^k\right\}$ be its structure constants over this basis and $\{x_1, .., x_n\}$ the corresponding dual basis. We define the following differential operators in the space $C^\infty(\mathfrak{g}^*)$:

$$\varphi(X_i) = \widehat{X}_i := C_{ij}^k x_k \frac{\partial}{\partial x_j}, \tag{3.1}$$

where $[X_i, X_j] = C_{ij}^k X_k$ $(1 \leq i < j \leq n)$. It is not difficult to verify that the operators $\widehat{X}_i$ satisfy the brackets $\left[\widehat{X}_i, \widehat{X}_j\right] = C_{ij}^k \widehat{X}_k$, therefore φ defines a representation of $\mathfrak{g}$. Using this fact, the invariance condition is then translated to an analytical condition [42]: Let $f = \sum_{i=1}^n f_i x_i$ be a linear function in $\mathfrak{g}^*$, $\xi \in \mathfrak{g}$ an arbitrary element and Exp: $\mathfrak{g} \to G$ the exponential function.

Proposition 2 *For an analytical function $F \in C^\infty(\mathfrak{g}^*)$ the following identity holds:*

$$\left. \frac{d^n}{dt^n} \right|_{t=0} F\left(Ad^*_{Exp(t\xi)} f\right) = (-\varphi^n(\xi) F)(f). \tag{3.2}$$

Proof. For any $\{x_1, ..., x_n\}$ we have the identity

$$x_i \left(Ad^*_{Exp(t\xi)} f\right) = \left(Ad^*_{Exp(t\xi)} f\right)(X_i) = f\left(Ad_{Exp(-t\xi)} X_i\right).$$

This implies that

$$\left. \frac{d}{dt} \right|_{t=0} F\left(Ad^*_{Exp(t\xi)} f\right) = \frac{\partial F}{\partial x_i}(f) \left. \frac{d}{dt} \right|_{t=0} x_i \left(Ad^*_{Exp(t\xi)} f\right)$$

$$= \frac{\partial F}{\partial x_i}(f) \left. \frac{d}{dt} \right|_{t=0} f\left(Ad_{Exp(-t\xi)} X_i\right),$$

and f being a linear function:

$$\left. \frac{d}{dt} \right|_{t=0} f\left(Ad_{Exp(-t\xi)} X_i\right) = f(-ad(\xi)(X_i)) = -C_{ji}^k \xi^j f_k. \tag{3.3}$$

Therefore

$$\left. \frac{d}{dt} \right|_{t=0} F\left(Ad^*_{Exp(t\xi)} f\right) = -C_{ji}^k \xi^j f_k \frac{\partial F}{\partial x_i}(\xi) = -(\varphi(\xi) F)(f).$$

Using the fact that

$$Ad^*_{Exp(t\xi)} Ad^*_{Exp(t'\xi)} f = Ad^*_{Exp((t+t')\xi)} f,$$

the general case $n > 1$ is proved recursively. $\blacksquare$

Theorem 1 *A function $F \in C^\infty(\mathfrak{g}^*)$ is a semi-invariant if and only if it is a solution of the following system:*

$$\widehat{X}_i F(x_1,..,x_n) = C_{ij}^k x_k \frac{\partial}{\partial x_j} F(x_1,..,x_n) = -d\chi(X_i) F, \qquad (3.4)$$

where $1 \le i \le n$ and $d\chi$ is the derivative of the character χ at the identity element of G. In particular, F is an invariant if and only if

$$\widehat{X}_i F = 0, \ 1 \le i \le n. \qquad (3.5)$$

Proof. From the semi-invariance condition

$$F\left(Ad^*_{Exp(t\xi)} f\right) = \chi(Exp(t\xi)) F(f)$$

it follows by (3.2) that

$$\left.\frac{d}{dt}\right|_{t=0} F\left(Ad^*_{Exp(t\xi)} f\right) = \left.\frac{d}{dt}\right|_{t=0} \chi(Exp(t\xi)) F(f) = -d\chi(\xi) F(f). \qquad (3.6)$$

In particular, if $\chi = 1$, equation (3.6) implies that

$$\widehat{X}_i(F) = 0, \ 1 \le i \le n.$$

Conversely, if $((-\varphi(\xi))^n F)(f) = d\chi(\xi) F(f)$, expanding $F\left(Ad^*_{Exp(t\xi)} f\right)$ as a Taylor series we obtain

$$F\left(Ad^*_{Exp(t\xi)} f\right) = F(f) + \sum_{n=1}^{\infty} \frac{(-\varphi(\xi))^n F}{n!}(f) t^n.$$

Inserting $d\chi(\xi) F(f)$ into the previous equation we obtain

$$F\left(Ad^*_{Exp(t\xi)} f\right) = F(f) + \sum_{n=1}^{\infty} \frac{(d\chi(\xi)) F}{n!}(f) t^n,$$

and since $\chi(Exp(t\xi)) = Exp(t\, d\chi(\xi))$ we deduce

$$F\left(Ad^*_{Exp(t\xi)} f\right) = \chi(Exp(t\xi)) F(f).$$

For the special case $\chi = 1$ we recover the invariance condition. $\blacksquare$

This theorem reduces the determination of the invariants to a system of linear first-order partial differential equations. A maximal set of functionally independent invariants of $\mathfrak{g}$ is called a fundamental set of invariants. For any given Lie algebra $\mathfrak{g}$, the number of functionally independent invariants can be computed from the brackets. More specifically, let $\left(C_{ij}^k x_k \right)$ be the matrix which represents the commutator table over the basis $\{X_1, .., X_n\}$, the C_{ij}^k being the structure constants over this basis. It follows from system (3.5) that the cardinal $\mathcal{N}(\mathfrak{g})$ of a fundamental set of invariants of $\mathfrak{g}$ is given by [7]:

$$\mathcal{N}(\mathfrak{g}) = \dim \mathfrak{g} - \mathrm{rank}\left(C_{ij}^k x_k \right). \tag{3.7}$$

In particular, the number of functionally independent invariants has the same parity as the dimension of the algebra, the commutator matrix being skew-symmetric. This number can also be derived using the equivalent approach of differential forms [15]. However, the number of polynomial solutions of (2) will be strictly lower than $\dim \mathfrak{g} - \mathrm{rank}\left(C_{ij}^k x_k \right)$, and only for particular classes of Lie algebras, such as semi-simple or nilpotent Lie algebras we will obtain an equality [1, 37]. As known, if $\mathfrak{A}$ denotes the universal enveloping algebra of $\mathfrak{g}$ and $Z(\mathfrak{A})$ its center, then the elements in $Z(\mathfrak{A})$ correspond to the Casimir operators [18]. This set indeed coincides with the set of polynomial invariants of Ad^*, while the rational invariants correspond to ratios of polynomials contained in $\mathfrak{A}$. The Casimir operators are recovered from system (3.5) using the symmetrization map

$$Sym\left(x_1^{a_1}, .., x_k^{a_k}\right) = \frac{1}{k!} \sum_{\sigma \in S_n} x_{\sigma(1)}^{a_1} .. x_{\sigma(k)}^{a_k},$$

S_k being the symmetric group on k letters [1]. However, the system (3.5) can also have non-rational solutions, which do not have such an interpretation in terms of the enveloping algebra. Such solutions are usually called generalized Casimir invariants.

Unfortunately, there is no general method to solve the system (3.5), and only for certain types of algebras fundamental systems of solutions can be determined. Recently an alternative method was proposed in [31], which turns out to be of interest for Lie algebras admitting only one (independent) invariant. This method reduces the problem of system (3.5) to the integration of a total differential equation whose coefficients can be computed by determinants. The system can thus be reduced to an equation:

$$dF = dx_1 + U_{12}dx_2 + \ldots + U_{1n}dx_n = 0, \tag{3.8}$$

where the U_{1i} are functions of the generators of $\mathfrak{g}$ obtained from the ratios $\frac{\partial F/\partial x_i}{\partial F/\partial x_1}$ by the rule of Cramer applied to the commutator matrix $A(\mathfrak{g})$ after having deleted the dependent row [31]. That is,

$$U_{1i} = \frac{\Delta_i}{\Delta_1} = \frac{f_i}{f_1} \tag{3.9}$$

where the Δ_i are simply the determinants obtained from the equation

$$A(\mathfrak{g})(x_i)^T = B, \tag{3.10}$$

the latter $(n \times 1)$-matrix expressing the coefficients of the dependent row. Therefore the solution of (2) is $F = \sum_{i=1}^{n} f_i x_i$, where f_i is the result of (5) after deleting the common factors of the determinants. Thus after evaluating $\dim(\mathfrak{g}) - 1$ determinants and integrating equation (4), the Casimir operator is found. For some specific types of Lie algebras having one invariant other direct methods to integrate the system have been developed [13].

If the Lie algebra is decomposable, i.e., $\mathfrak{g} = \mathfrak{s} \oplus \mathfrak{r}$, then it is a direct consequence of that $\mathcal{N}(\mathfrak{g}) = \mathcal{N}(\mathfrak{s}) + \mathcal{N}(\mathfrak{r})$. Since the sum is direct, we have that $[\mathfrak{s}, \mathfrak{r}] = 0$ and therefore the rank of the matrix $A(\mathfrak{g})$ is the sum of the ranks of $A(\mathfrak{s})$ and $A(\mathfrak{r})$. For algebras having a non-trivial Levi decomposition, no apparent relation between the number of invariants of the Levi factor and the radical and the number of invariants of the semidirect sum exists, as shown by the special affine Lie algebras [13, 31]. In particular, it is not sufficient to determine the invariants of solvable Lie algebras to have an overview of invariants of Lie algebras. At the present time there is no hope to construct a theory of invariants that covers all types of Lie algebras.

Chapter 4

Invariants of Nine Dimensional Real Algebras

In this section we recall some specific results and properties that have been used to solve system (3.5) for the nine dimensional Lie algebras with nontrivial Levi decomposition. To this extent, we recall the notions of contraction of Lie algebras (for general properties see e.g. [47]). Let $\mathfrak{g}$ be a Lie algebra and $\Phi_t \in Aut(\mathfrak{g})$ a family of nonsingular linear maps of $\mathfrak{g}$, where $t \in \mathbb{N}$. For any $X, Y \in \mathfrak{g}$ define

$$[X,Y]_{\Phi_t} := [\Phi_t(X), \Phi_t(Y)]. \tag{4.1}$$

Obviously $[X,Y]_{\Phi_t}$ are the brackets of the Lie algebra over the transformed basis. If the limit

$$[X,Y]_\infty := \lim_{t \to \infty} \Phi_t^{-1}[\Phi_t(X), \Phi_t(Y)] \tag{4.2}$$

exists for any $X, Y \in \mathfrak{g}$, equation (4.2) defines a Lie algebra $\mathfrak{g}'$ called contraction of $\mathfrak{g}$, nontrivial if $\mathfrak{g}$ and $\mathfrak{g}'$ are non-isomorphic.

Lemma 1 *Let $\mathfrak{g}$ be a nine dimensional indecomposable Lie algebra with nontrivial Levi decomposition. Then $\mathcal{N}(\mathfrak{g}) = 1, 3$.*

Proof. Let $\{X_1, .., X_n, X_{n+1}, .., X_{n+m}\}$ be a basis of $\mathfrak{g}$ such that $\{X_1, .., X_n\}$ is a basis of $\mathfrak{s}$ and $\{X_{n+1}, .., X_{n+m}\}$ is a basis of $\mathfrak{r}$. Let $\left\{C_{ij}^k\right\}_{1 \le i,j,k \le n+m}$ be the structure constants of $\mathfrak{g}$ over this basis. If we consider the change of basis defined by:

$$X'_i := X_i, \ 1 \le i \le n,$$
$$X'_{n+j} := \tfrac{1}{n} X_{n+j}, \ 1 \le j \le m, \tag{4.3}$$

then, over the new basis the brackets are:

$$\left[X'_i, X'_j \right] = [X_i, X_j], \ 1 \le i, j \le n$$
$$\left[X'_i, X'_{n+j} \right] = \tfrac{1}{n}[X_i, X_{n+j}], \ 1 \le i \le n, 1 \le j \le m \tag{4.4}$$
$$\left[X'_{n+i}, X'_{n+j} \right] = \tfrac{1}{n^2}[X_{n+i}, X_{n+j}], \ 1 \le i, j \le m.$$

Since $\left[X'_i, X'_{n+j} \right] = \tfrac{1}{n} C^{n+k}_{i,j+n} X_{n+k} = C^{n+k}_{i,j+n} X'_{n+k}$, this shows that the Levi part $\mathfrak{s}$ and the representation R of $\mathfrak{s}$ on the radical remain unchanged, while the brackets of the radical $\mathfrak{r}$ adopt the form:

$$[X'_{n+i}, X'_{n+j}] = \frac{1}{n} C^{n+k}_{n+i,n+j} X'_{n+k}, \ 1 \le i, j, k \le m. \tag{4.5}$$

If we consider the limit $n \to \infty$, we obtain a Lie algebra with the same Levi subalgebra and describing representation, but where the radical satisfies the brackets

$$[X'_{n+i}, X'_{n+j}] = 0.$$

This shows that the affine Lie algebra $\mathfrak{s} \overrightarrow{\oplus}_R (\deg R) L_1$ is a Inönü-Wigner contraction of $\mathfrak{g}$. Using either the fact that contractions increase or leave invariant the number of independent solutions of (3.5), it follows that

$$\mathcal{N}(\mathfrak{g}) \le \mathcal{N}\left(\mathfrak{s} \overrightarrow{\oplus}_R (\deg R) L_1 \right). \tag{4.6}$$

Table 1 indicates the number of invariants of nine dimensional affine algebras obtained by contraction of indecomposable algebras $\mathfrak{g}$ having a non-trivial Levi decomposition. The problem is therefore reduced to prove that any indecomposable Lie algebra contracting onto $\mathfrak{so}(3) \overrightarrow{\oplus}_{ad\mathfrak{so}(3)\oplus 3D_0} 6L_1$ or $\mathfrak{sl}(2,\mathbb{R}) \overrightarrow{\oplus}_{D_{\frac{1}{2}}\oplus 4D_0} 6L_1$ has at most three invariants[1]. We prove it for the first algebra, the remaining case being analogous.

---

[1] Since $\left(\mathfrak{sl}(2,\mathbb{R}) \overrightarrow{\oplus}_{D_1 \oplus 3D_0} 6L_1 \right) \otimes_{\mathbb{R}} \mathbb{C}$ is isomorphic to $\left(\mathfrak{so}(3) \overrightarrow{\oplus}_{ad\mathfrak{so}(3)\oplus 3D_0} 6L_1 \right) \otimes_{\mathbb{R}} \mathbb{C}$, it suffices to prove the assertion for one of the algebras, the property being preserved by the real forms.

Let $\{X_4,..,X_9\}$ be a basis of the radical $\mathfrak{r}$ such that $\{X_4,..,X_6\}$ transforms according to the adjoint representation of $\mathfrak{so}(3)$ and the remaining elements by the trivial representation. The Jacobi conditions imply that

$$[X_4,X_5] = [X_4,X_6] = [X_5,X_6] = 0,$$
$$[X_i,X_{7+j}] = \lambda_j X_i, \ i = 4,5,6; \ j = 0,1,2,$$
$$[X_{7+i},X_{7+j}] \in \mathbb{R}\langle X_7,X_8,X_9\rangle, \ 0 \le i < j \le 2.$$

Since the algebra is indecomposable, at least one λ_i must be nonzero. Without loss of generality we can suppose that $\lambda_0 \ne 0$. A change of basis allows to put $\lambda_1 = \lambda_2 = 0$. In order that $\mathfrak{g}$ is indecomposable, the following case holds:

$$[X_7,X_8] = \alpha_1 X_8 + \beta X_9,$$
$$[X_7,X_9] = \gamma X_8 + \alpha_2 X_9$$

with $\mathrm{rank}\begin{pmatrix} \alpha_1 & \beta \\ \gamma & \alpha_2 \end{pmatrix} \ge 1$. Further we obtain that

$$[X_8,X_9] = \delta_1 X_8 + \delta_2 X_9$$

with

$$\alpha_2\delta_1 - \delta_2\gamma = \alpha_1\delta_2 - \delta_1\beta = 0.$$

In any case it follows that

$$\begin{pmatrix} 0 & \alpha_1 x_8 + \beta x_9 & \gamma x_8 + \alpha_2 x_9 \\ -\alpha_1 x_8 - \beta x_9 & 0 & \delta_1 x_8 + \delta_2 x_9 \\ -\gamma x_8 - \alpha_2 x_9 & -\delta_1 x_8 - \delta_2 x_9 & 0 \end{pmatrix} \ge 2,$$

showing that

$$\mathrm{rank}\,A\,(\mathfrak{g}) \ge 6,$$

and therefore $\mathcal{N}(\mathfrak{g}) \le 3$. $\blacksquare$

We now recall some properties that have been used in the integration of the system (3.5) for the specific case of nine dimensional algebras with Levi decomposition.

Proposition 3 *If $\mathfrak{g}$ admits a codimension one subalgebra $\mathfrak{k}$, then $\mathfrak{g}$ and $\mathfrak{k}$ have $\left[\frac{\mathcal{N}(\mathfrak{g})}{2}\right]$ invariants in common.*

R. Campoamor-Stursberg

Table 4.1. Affine Lie algebras obtained by contraction from nine dimensional algebras with Levi part $\mathfrak{s} \neq 0$.

Affine Lie algebra $\mathfrak{g}$	$\mathcal{N}(\mathfrak{g})$	
$\mathfrak{so}(3) \overrightarrow{\oplus}_{ad\mathfrak{so}(3)\oplus 3D_0} 6L_1$	5	decomposable
$\mathfrak{so}(3) \overrightarrow{\oplus}_{R_4\oplus 2D_0} 6L_1$	3	decomposable
$\mathfrak{so}(3) \overrightarrow{\oplus}_{R_5\oplus D_0} 6L_1$	3	decomposable
$\mathfrak{so}(3) \overrightarrow{\oplus}_{2ad\mathfrak{so}(3)} 6L_1$	3	indecomposable
$\mathfrak{sl}(2,\mathbb{R}) \overrightarrow{\oplus}_{D_{\frac{1}{2}}\oplus 4D_0} 6L_1$	5	decomposable
$\mathfrak{sl}(2,\mathbb{R}) \overrightarrow{\oplus}_{D_1\oplus 3D_0} 6L_1$	5	decomposable
$\mathfrak{sl}(2,\mathbb{R}) \overrightarrow{\oplus}_{D_{\frac{3}{2}}\oplus 2D_0} 6L_1$	3	decomposable
$\mathfrak{sl}(2,\mathbb{R}) \overrightarrow{\oplus}_{2D_{\frac{1}{2}}\oplus 2D_0} 6L_1$	3	decomposable
$\mathfrak{sl}(2,\mathbb{R}) \overrightarrow{\oplus}_{D_1\oplus D_{\frac{1}{2}}\oplus D_0} 6L_1$	3	decomposable
$\mathfrak{sl}(2,\mathbb{R}) \overrightarrow{\oplus}_{D_2\oplus D_0} 6L_1$	3	decomposable
$\mathfrak{sl}(2,\mathbb{R}) \overrightarrow{\oplus}_{D_{\frac{5}{2}}} 6L_1$	3	indecomposable
$\mathfrak{sl}(2,\mathbb{R}) \overrightarrow{\oplus}_{D_{\frac{3}{2}}\oplus D_{\frac{1}{2}}} 6L_1$	3	indecomposable
$\mathfrak{sl}(2,\mathbb{R}) \overrightarrow{\oplus}_{2D_1} 6L_1$	3	indecomposable
$\mathfrak{sl}(2,\mathbb{R}) \overrightarrow{\oplus}_{3D_{\frac{1}{2}}} 6L_1$	3	indecomposable

Proof. In [33] it was shown that the number of invariants of a Lie algebra $\mathfrak{g}$ is related with that of subalgebras by means of the following inequality:

$$\dim \mathfrak{g} - \mathcal{N}(\mathfrak{g}) - \dim \mathfrak{k} - \mathcal{N}(k) + 2l' \geq 0,$$

where l' is the number of invariants of $\mathfrak{g}$ depending only on variables of the subalgebra $\mathfrak{k}$. For the case of codimension one subalgebras we therefore obtain

$$1 - \mathcal{N}(\mathfrak{g}) - \mathcal{N}(k) + 2l' \geq 0.$$

By the preceding lemma, two cases must be distinguished:

1. If $\mathcal{N}(\mathfrak{g}) = 3$, we obtain the inequality

$$2l' - \mathcal{N}(k) - 2 \geq 0,$$

   thus

$$l' \geq 1 + \frac{\mathcal{N}(\mathfrak{k})}{2} \geq \left[\frac{\mathcal{N}(\mathfrak{g})}{2} \right].$$

   In particular, in this case the codimension subalgebra can possess up to four independent invariants.

2. If $\mathcal{N}(\mathfrak{g}) = 1$, then

$$2l' - \mathcal{N}(\mathfrak{k}) \geq 0,$$

   and since

$$2 \geq 2l' \geq \mathcal{N}(k),$$

   either $l' = 0$ or $l' = 1$. The subalgebra $\mathfrak{k}$ has at most two invariants.

∎

The preceding result is of interest for the computation of invariants using codimension one subalgebras of $\mathfrak{g}$ (if they exist). More specifically, suppose that $\mathfrak{k}$ is a codimension one subalgebra of $\mathfrak{g}$ and $\mathcal{F} = \{J_1,..,J_l\}$ a fundamental system of invariants of $\mathfrak{k}$. Suppose further that there exists an element $X \in \mathfrak{g}$ with $X \notin [\mathfrak{g},\mathfrak{g}]$ such that for any invariant F of $\mathfrak{g}$ we have $\frac{\partial F}{\partial x} = 0$. Then the system (3.5) associated to $\mathfrak{g}$ is the union of the subsystem corresponding to the subalgebra $\mathfrak{k}$ and the equation

$$\widehat{X}(F) = C_{iX}^{j} x_{j} \frac{\partial F}{\partial x_{i}} = 0. \tag{4.7}$$

If the set $\mathcal{F} = \{J_1,..,J_l\}$ transforms linearly by the differential operator $\widehat{X}$, i.e., if

$$\widehat{X}(J_i) = \alpha^{j} J_{j}, \ 1 \leq i \leq l,$$

then we can always find a function $\Phi(J_1,..,J_l)$ that is a common solution to the subalgebra and the equation (4.7) [21]. This procedure always works when we have a Lie algebra that is an extension by a derivation of a codimension one subalgebra [13], and constitutes the main technique employed in this work to compute the invariants of nine dimensional Lie algebras.

As example, consider the Lie algebra $L_{9,7}^{*}$ given by the brackets (2.3). In this case, the system to be solved is

$$\begin{pmatrix} 0 & x_3 & -x_2 & \frac{x_7}{2} & \frac{x_6}{2} & -\frac{x_5}{2} & -\frac{x_4}{2} & 0 & 0 \\ -x_3 & 0 & x_1 & \frac{x_5}{2} & -\frac{x_4}{2} & \frac{x_7}{2} & -\frac{x_6}{2} & 0 & 0 \\ x_2 & -x_1 & 0 & \frac{x_6}{2} & -\frac{x_7}{2} & -\frac{x_4}{2} & \frac{x_5}{2} & 0 & 0 \\ -\frac{x_7}{2} & -\frac{x_5}{2} & -\frac{x_6}{2} & 0 & 0 & x_8 & 0 & 0 & x_4 \\ -\frac{x_4}{2} & \frac{x_4}{2} & \frac{x_7}{2} & 0 & 0 & 0 & x_8 & 0 & x_5 \\ \frac{x_5}{2} & -\frac{x_7}{2} & \frac{x_4}{2} & -x_8 & 0 & 0 & 0 & 0 & x_6 \\ \frac{x_4}{2} & \frac{x_6}{2} & -\frac{x_5}{2} & 0 & -x_8 & 0 & 0 & 0 & x_7 \\ 0 & 0 & 0 & 0 & 0 & 0 & 0 & 0 & 2x_8 \\ 0 & 0 & 0 & -x_4 & -x_5 & -x_6 & -x_7 & -2x_8 & 0 \end{pmatrix} \begin{pmatrix} \partial_{x_1} F \\ \partial_{x_2} F \\ \partial_{x_3} F \\ \partial_{x_4} F \\ \partial_{x_5} F \\ \partial_{x_6} F \\ \partial_{x_7} F \\ \partial_{x_8} F \\ \partial_{x_9} F \end{pmatrix} = 0$$

Since the rank of the matrix is 8, there is only one invariant. It follows from

$$\widehat{X}_8(F) = 2x_8 \frac{\partial F}{\partial x_9} = 0$$

that the solutions do not depend on x_9. We can therefore consider the eight dimensional subalgebra of $L_{9,7}^{*}$ spanned by $\langle X_1,..,X_8 \rangle$. This subalgebra is isomorphic to the semidirect product of $\mathfrak{so}(3)$ with a five dimensional Heisenberg

Lie algebra $\mathfrak{h}_2$, and a fundamental system of invariants can be computed by by means of determinantal methods [16]. By this procedure we obtain

$$I_1 = 16\left(x_1^2 + x_2^2 + x_3^2\right)x_8^2 + \left(x_4^2 + x_5^2 + x_6^2 + x_7^2\right)^2 + 16x_1x_8\left(x_4x_5 + x_6x_7\right) +$$
$$- 16x_2x_8\left(x_5x_6 - x_4x_7\right) + 8x_3x_8\left(x_4^2 - x_5^2 + x_6^2 - x_7^2\right),$$
$$I_2 = x_8.$$

If we evaluate these functions in the differential operator $\widehat{X}_9$, it is straightforward to verify that

$$\widehat{X}_9\left(I_1\right) = -4I_1,$$
$$\widehat{X}_9\left(I_2\right) = -2I_2. \tag{4.8}$$

We will call (4.8) the semi-invariance conditions with respect to the operator $\widehat{X}_9$. Since I_1 and I_2 are semi-invariants of $L_{9,7}^*$, in order to find an invariant, we consider the new variables $u = I_1$ and $v = I_2$ and the differential equation

$$\frac{\partial}{\partial u}F\left(u,v\right) + \frac{\widehat{X}_9\left(v\right)}{\widehat{X}_9\left(u\right)}\frac{\partial}{\partial v}F\left(u,v\right) = 0, \tag{4.9}$$

with solution $J = uv^{-2}$. Therefore the function $J = I_1I_2^{-2}$ provides the invariant of $L_{9,7}^*$. It can actually be shown that $L_{9,7}^*$ is an extension of the subalgebra $L\langle X_1,..,X_8\rangle$ by a derivation of it radical (isomorphic to $\mathfrak{h}_2$).

Indecomposable Lie algebras with a Levi subalgebra of rank one have some special properties that do not hold for higher ranks, and that are useful for the computation of invariants [12]. These often allow to find adequate subalgebras to obtain the semi-invariants of the algebra, or even to obtain alternative criteria to integrate system (3.5):

Proposition 4 *Let* $\mathfrak{s} = \mathfrak{so}\left(3\right),\mathfrak{sl}\left(2,\mathbb{R}\right)$. *If the radical* $\mathfrak{r}$ *of* $\mathfrak{s}\overrightarrow{\oplus}_R\mathfrak{r}$ *has a one dimensional centre, then the representation R describing the semidirect sum contains a copy of the trivial representation* D_0.

This result specially applies to nilpotent radicals that contract onto the Heisenberg algebra or solvable radicals that contain some zero weight for its action on the nilradical [3, 41].

We finally recall a result that is of interest for Lie algebras with abelian radical.

Proposition 5 *Let R be a representation of $\mathfrak{s}$, where $\mathfrak{s} = \mathfrak{so}(3)$ or $\mathfrak{sl}(2,\mathbb{R})$. If $\dim R > 3$ and the radical of $\mathfrak{s} \overrightarrow{\oplus}_R \mathfrak{r}$ is abelian, then there exists a fundamental set of invariants formed by functions F_i depending only on variables associated to elements of $\mathfrak{r}$.*

A proof can be found in [12] or [27]. This case turns out to be usually more difficult to integrate, particularly when the representation R is irreducible. In the case that an algebra has an abelian radical, the contraction procedure is often employed to obtain the invariants [15, 22]. For algebras that cannot be obtained by a limiting process, we are often led to integrate the corresponding system (3.5) directly. Even if we can make some predictions on the invariants, the effective computation for some affine Lie algebras (i.e., semidirect product of abelian and semisimple Lie algebras) can be extremely complicated, due to the enormous number of terms (see e.g. the Lie algebra $L_{9,59}$). This seems to indicate that, even if the system (3.5) is extremely simple in appearance, no general method leading to a fundamental system of invariants can be obtained [17]. Recently, a new geometrical method has been introduced in [8, 9], which reduces system (3.5) to a purely algebraic system. This approach has been proven to be effective for various large types of solvable algebras in arbitrary dimension. It remains to see whether its implementation for the non-solvable case provides some additional simplification that could provide a complete integration of the above type of affine algebras. Work in this direction is in progress.

Chapter 5

Rigidity and Completeness

In this chapter we classify the rigid and complete Lie algebras of dimension nine with non-trivial Levi decomposition. This distinguished class of algebras is of importance for the general analysis, as well as for the classification of contractions, since rigid Lie algebras cannot appear as a contraction of a non-isomorphic Lie algebra [11]. To this extent, we have to recall some questions concerning the geometrical interpretation of Lie algebras and their cohomology. Since Lie algebras $\mathfrak{g}$ are determined by a tensor, they can be seen as a pair (V,μ), where V is a vector space endowed with a skew-symmetric tensor μ of type $(1,2)$ that satisfies the Jacobi identity. If $\mathcal{M}$ denotes the manifold consisting of all Lie algebra laws on V, then the linear group $GL(V,\mathbb{R})$ acts on the space of all alternating bilinear forms over V by

$$(f * \mu) = f^{-1}(\mu(f,f)), \quad f \in GL(V,\mathbb{R}), \tag{5.1}$$

showing that $\mathcal{M}$ is stable under this action. Therefore, the orbits $O(\mu)$ of the linear group $GL(V,\mathbb{R})$ on $\mathcal{M}$ correspond to the isomorphism classes of Lie algebra laws on V.

Definition 1 *A Lie algebra* $\mathfrak{g} = (V,\mu)$ *is called rigid if its orbit* $O(\mu)$ *is open in* $\mathfrak{M}$.

Roughly speaking, a Lie algebra $\mathfrak{g}$ is rigid if any Lie algebra $\mathfrak{g}'$ close to $\mathfrak{g}$ is isomorphic to $\mathfrak{g}$ [6, 10]. It turns out that rigidity can be studied using cohomology of Lie algebras, although it is known that rigid Lie algebras with non-vanishing cohomology exist. Let $\mathfrak{g}$ be a Lie algebra and V a representation of $\mathfrak{g}$.

A p-dimensional cochain of $\mathfrak{g}$ (with values in V) is a p-linear skew-symmetric mapping

$$\Phi : \mathfrak{g} \times .^{(p)}. \times \mathfrak{g} \longrightarrow V. \tag{5.2}$$

A 0-cochain is by definition a constant function. We denote by $C^p(\mathfrak{g},V) = Hom(\wedge^p \mathfrak{g},V)$ the space of p-cochains. We can provide $C^p(\mathfrak{g},V)$ with the structure of a $\mathfrak{g}$-module structure by putting

$$(X\Phi)(X_1,...,X_p) = X.\Phi(X_1,...,X_p) - \sum_{1\leq i\leq p} \Phi(X_1,...,[X,X_i],...,X_p) \tag{5.3}$$

for all $X_1,...,X_p \in \mathfrak{g}$. The coboundary operator $d_p : C^p(\mathfrak{g},V) \longrightarrow C^{p+1}(\mathfrak{g},V)$ is defined by

$$d_p\Phi(X_1,...,X_{p+1}) = \sum_{1\leq s\leq p+1}(-1)^{s+1}(X_s.\Phi)\left(X_1,...,\widehat{X_s},...,X_{p+1}\right) + $$
$$+\sum_{1\leq s\leq t\leq p+1}(-1)^{s+t}\Phi\left([X_s,X_t],X_1,...,\widehat{X_s},...,\widehat{X_t},...,X_{p+1}\right) \tag{5.4}$$

By this definition, $d_p(C^p(\mathfrak{g},\mathfrak{g})) \subset C^{p+1}(\mathfrak{g},\mathfrak{g})$, and it can be verified that $d_p \circ d_p = 0$ for all p. The space of p-cocycles is defined as $Z^p(\mathfrak{g},V) = \ker d_p$, and coboundaries by $B^p(\mathfrak{g},V) = \mathrm{Im}\, d_p$. The pth-cohomology space with values in V is then defined by

$$H^p(\mathfrak{g},V) = Z^p(\mathfrak{g},V)/B^p(\mathfrak{g},V). \tag{5.5}$$

In particular, for any $p \geq$ we have the following identity:

$$\dim B^{p+1}(\mathfrak{g},V) = \dim C^p(\mathfrak{g},V) - \dim Z^p(\mathfrak{g},V) = \dim V \binom{\dim \mathfrak{g}}{p} - \dim Z^p(\mathfrak{g},V). \tag{5.6}$$

For the trivial module $V = \mathbb{R}$ the notation for the cohomology spaces is simply $H^p(\mathfrak{g})$. We also recall the interpretation of some cohomology groups of low order (see e.g. [6] for further details):

1. $H^0(\mathfrak{g},\mathfrak{g}) = Z(\mathfrak{g})$ is the centre of $\mathfrak{g}$.

2. $H^1(\mathfrak{g},\mathfrak{g}) = ODer(\mathfrak{g})$ is the algebra of outer derivations.

3. $\dim H^1(\mathfrak{g}) = \mathrm{codim}_{\mathfrak{g}}[\mathfrak{g},\mathfrak{g}]$.

4. $H^2(\mathfrak{g},\mathfrak{g})$ is identified with thegenerators of infinitesimal deformations.

5. $H^2(\mathfrak{g})$ is identified with the isomorphism classes of one dimensional central extensions of $\mathfrak{g}$.

The relation between the geometrical structure of an open orbit $O(\mu)$ and the cohomology of the corresponding Lie algebra $\mathfrak{g}=(V,\mu)$ is given by the Nijenhuis-Richardson theorem [25]

Theorem 2 *Let $\mathfrak{g}$ be a (real) Lie algebra such that $H^2(\mathfrak{g},\mathfrak{g})=0$. Then $\mathfrak{g}$ is rigid.*

As commented, this result constitutes a sufficient, but not necessary condition, as pointed out by Nijenhuis and Richardson. A direct consequence of this fact is that semi-simple Lie algebras, which have vanishing second adjoint cohomology groups, are rigid. For semidirect products of semisimple and solvable Lie algebras, much less is known about their cohomology. There exists however a structure theorem that imposes some restrictions on algebras having this property:

Theorem 3 *Any rigid Lie algebra $\mathfrak{g}$ over $\mathbb{K}=\mathbb{R},\mathbb{C}$ satisfies one of the following conditions:*

1. *The radical $\mathfrak{r}$ is not nilpotent and satisfies $\dim Der(\mathfrak{g})=\dim\mathfrak{g}$. If moreover $codim_{\mathfrak{g}}[\mathfrak{g},\mathfrak{g}]>1$, then $\mathfrak{g}$ is complete.*

2. *The radical is nilpotent and verifies one of the following constraints:*

   (a) *$\mathfrak{g}$ is perfect (i.e., $\mathfrak{g}=[\mathfrak{g},\mathfrak{g}]$),*

   (b) *$\mathfrak{g}$ is the direct sum of $\mathbb{K}$ and one perfect rigid algebra, the derivations of which are all inner,*

   (c) *$\mathfrak{g}$ is not perfect, has no direct non-zero Abelian factor and satisfies $\mathfrak{t}_e(\mathfrak{g})=0$.*

It is not uncommon for rigid Lie algebras with vanishing $H^2(\mathfrak{g},\mathfrak{g})$ that other cohomology groups $H^i(\mathfrak{g},\mathfrak{g})$ are also zero. A special case is when $H^0(\mathfrak{g},\mathfrak{g})=H^1(\mathfrak{g},\mathfrak{g})=0$. Such algebras are called complete, and from the previous remarks it follows that such an algebra has no center and no outer derivations. For the case of non-trivial Levi decompositions, the possibilities of having at most $\dim\mathfrak{g}$ derivations are resumed in the following Table:

**Table 5.1. Structural constraints for Lie algebras $\mathfrak{g}$
satisfying the condition $\dim\mathfrak{g} = \dim Der(\mathfrak{g})$**

$\dim Der(\mathfrak{g})$	$Z(\mathfrak{g})$	$\mathrm{codim}_{\mathfrak{g}}\,[\mathfrak{g},\mathfrak{g}]$	$H^1(\mathfrak{g},\mathfrak{g})$	Existence
$\dim\mathfrak{g}$	0	0	0	yes
$\dim\mathfrak{g}$	0	≥ 1	0	yes
$\dim\mathfrak{g}$	$\neq 0$	0	$\neq 0$	yes
$\dim\mathfrak{g}$	$\neq 0$	1	$\neq 0$	yes

Following the exhaustive analysis of nine dimensional algebras in the next chapter, rigid and complete algebras are easily classified.

Proposition 6 *Let $\mathfrak{g}$ be an indecomposable nine dimensional real Lie algebra with non-trivial Levi decomposition. Then $\mathfrak{g}$ is rigid if and only if it is isomorphic to one of the following algebras:*

$$L_{9,4}, L_{9,8}, L_{9,9}, L_{9,36}, L_{9,37}, L_{9,38}, L_{9,42}, L_{9,50}, L_{9,54}, L_{9,55}, L_{9,59}, L_{9,60}, L_{9,63}.$$

The proof follows from the computation of the cohomology group $H^2(\mathfrak{g},\mathfrak{g})$ for the Lie algebras $L_{9,i}$. For those Lie algebras depending on parameters, it is obvious that no rigidity is possible, since any change of parameter defines a non-zero cocycle, while for the remaining non-trivial deformations are easily constructed. Thus a nine dimensional Lie algebra $\mathfrak{s}\,\overrightarrow{\oplus}_R\,\mathfrak{r}$ is rigid if and only if it satisfies the Nijenhuis-Richardson theorem. As for completeness of these algebras, the classification is resumed in the next

Proposition 7 *Let $\mathfrak{g}$ be an indecomposable nine dimensional real Lie algebra with non-trivial Levi decomposition. Then $\mathfrak{g}$ is complete if and only if it is isomorphic to one of the following algebras:*
$$L_{9,8}, L_{9,9}, L_{9,36}, L_{9,38}, L_{9,50}, L_{9,54}, L_{9,55}.$$

By inspection, it follows at once that any of the complete algebras above is also rigid, although the converse fails, as happens for $L_{9,63}$. In particular, the algebras $L_{9,4}, L_{9,37}$ and $L_{9,42}$ have a nine dimensional derivation algebra, but a non-trivial center. We therefore find examples for all possibilities of Table 5.1.

Chapter 6

Tables of Invariants of Nine Dimensional Algebras

We shortly resume the notation used along the tables. Let $\mathfrak{g} = \mathfrak{s} \overrightarrow{\oplus}_R \mathfrak{r}$ be a nine dimensional Lie algebra with nontrivial Levi decomposition, where $\mathfrak{s}$ is the Levi subalgebra, $\mathfrak{r}$ the radical and R the representation of $\mathfrak{s}$ describing the semidirect product. The structure tensor $\left\{ C_{ij}^k \right\}$ of $\mathfrak{g}$ is given over a basis $\{X_1, .., X_9\}$, where $\{X_1, X_2, X_3\}$ is a basis of the Levi subalgebra and $\{X_4, .., X_9\}$ is the basis of the radical. The following information is given in the tables:

- The Levi decomposition and the structure tensor after the classification in [45]. The subindex k in the notation $L_{9,k}^{p_1,...,p_r}$ indicates the number of the isomorphism class, while super-indices $p_1, .., p_k$ refer to continuous parameters (if any).

- The notation for the radicals are those employed in references [24] and [29].

- The codimension of the derived ideal $[\mathfrak{g}, \mathfrak{g}]$.

- The dimension of the Lie algebra of derivations $Der(\mathfrak{g})$.

- The dimension of the orbit $O(\mathfrak{g})$ of $\mathfrak{g}$.

- The rank of the matrix $A(\mathfrak{g}) = \left(C_{ij}^k x_k \right)$.

- Conditions on the invariants of the algebra. If the invariants depend on all variables $\{x_1,..,x_9\}$, then the invariants are searched either by direct integration or using alternative methods.

- The subalgebra $L\langle X_{i_1},..,X_{i_k}\rangle$ of $\mathfrak{g}$ generated by the elements $\{X_{i_1},..,X_{i_k}\}$ providing semi-invariants of $\mathfrak{g}$, only in the case that the invariants of $\mathfrak{g}$ can be expressed in terms of some subalgebra invariants. The isomorphism class uses the notations of [44]

- A fundamental system of invariants $\{J_1,..,J_t\}$ of $L\langle X_{i_1},..,X_{i_k}\rangle$.

- The semi-invariance conditions.

- Functions $\Phi_l(J_1,..,J_t)$ providing a fundamental system of invariants of $\mathfrak{g}$.

$\mathbf{L}_{9,1}^{p,q}$ $[pq \neq 0]$

- Levi decomposition: $L_{9,1}^{p,q} = \mathfrak{so}(3)\,\overrightarrow{\oplus}\, R\mathfrak{g}_{6,1}$

- Describing representation: $R = ad\mathfrak{so}(3)\oplus 3D_0$.

- Structure tensor

$$
\begin{array}{llllll}
C_{12}^3 = 1, & C_{13}^2 = -1, & C_{23}^1 = 1, & C_{15}^6 = 1, & C_{16}^5 = -1, & C_{24}^6 = -1,\\
C_{26}^4 = 1, & C_{34}^5 = 1, & C_{35}^4 = -1, & C_{49}^4 = 1, & C_{59}^5 = 1, & C_{69}^6 = 1,\\
C_{79}^7 = p, & C_{89}^8 = q. & & & &
\end{array}
$$

- $\mathrm{codim}_{\mathfrak{g}}[\mathfrak{g},\mathfrak{g}] = 1$.

- $\dim Der(\mathfrak{g}) = \begin{cases} 11, & p \neq q \\ 13, & p = q \end{cases}$.

- $\dim O(\mathfrak{g}) = \begin{cases} 70, & p \neq q \\ 68, & p = q \end{cases}$.

- Rank $A(\mathfrak{g}) = 6$.

- Conditions on invariants:

$$
\frac{\partial F}{\partial x_9} = 0.
$$

- Codimension one subalgebra: $L\langle X_1, .., X_8\rangle \simeq L_{6,1} \oplus 2L_1$

- Invariants of subalgebra:

$$J_1 = x_4^2 + x_5^2 + x_6^2$$
$$J_2 = x_1 x_4 + x_2 x_5 + x_3 x_6$$
$$J_3 = x_7, \; J_4 = x_8$$

- Semi-invariance conditions:

$$\widehat{X}_9(J_1) = -2J_1, \; \widehat{X}_9(J_2) = -J_2$$
$$\widehat{X}_9(J_3) = -pJ_3, \; \widehat{X}_9(J_4) = -qJ_4.$$

- Invariants of $L_{9,1}^{p,q}$

$$I_1 = \frac{J_1^p}{J_3^2}, \; I_2 = \frac{J_2^p}{J_3}, \; I_3 = \frac{J_3^q}{J_4^p}.$$

$\mathbf{L}_{9,2}^p$

- Levi decomposition: $L_{9,2}^p = \mathfrak{so}(3) \overrightarrow{\oplus}_R \mathfrak{g}_{6,2}$

- Describing representation: $R = ad\mathfrak{so}(3) \oplus 3D_0$.

- Structure tensor

$$C_{12}^3 = 1, \quad C_{13}^2 = -1, \quad C_{23}^1 = 1, \quad C_{15}^6 = 1, \quad C_{16}^5 = -1, \quad C_{24}^6 = -1,$$
$$C_{26}^4 = 1, \quad C_{34}^5 = 1, \quad C_{35}^4 = 1, \quad C_{49}^4 = 1, \quad C_{59}^5 = 1, \quad C_{69}^6 = 1,$$
$$C_{79}^7 = p, \quad C_{89}^8 = 1, \quad C_{79}^8 = p.$$

- $\operatorname{codim}_{\mathfrak{g}}[\mathfrak{g},\mathfrak{g}] = \begin{cases} 1, & p \neq 0 \\ 2, & p = 0 \end{cases}.$

- $\dim Der(\mathfrak{g}) = 11$.

- $\dim O(\mathfrak{g}) = 70$.

- Rank $A(\mathfrak{g}) = 6$

- Conditions on invariants:

$$\frac{\partial F}{\partial x_9} = 0.$$

- Codimension one subalgebra: $L\langle X_1, .., X_8 \rangle \simeq L_{6,1} \oplus 2L_1$

- Invariants of subalgebra:

$$J_1 = x_4^2 + x_5^2 + x_6^2$$
$$J_2 = x_1 x_4 + x_2 x_5 + x_3 x_6$$
$$J_3 = x_7, \; J_4 = x_8$$

- Semi-invariance conditions:

$$\widehat{X}_9(J_1) = -2J_1, \; \widehat{X}_9(J_2) = -J_2$$
$$\widehat{X}_9(J_3) = -pJ_3, \; \widehat{X}_9(J_4) = -J_3 - pJ_4.$$

- Invariants of $L_{9,2}^p$

$$I_1 = \frac{J_1}{J_2^2}, \; I_2 = \frac{J_2^p}{J_3}, \; I_3 = -\frac{J_3}{J_4} + \frac{1}{p}\ln(J_3), \; p \neq 0.$$
$$I_1 = \frac{J_1}{J_2^2}, \; I_2 = J_3, \; I_3 = J_1 \exp(-J_4/J_3), \; p = 0.$$

$\mathbf{L}_{9,3}^{p,q} \; [p \neq 0, \; q \geq 0]$

- Levi decomposition: $L_{9,3}^{p,q} = \mathfrak{so}(3) \overrightarrow{\oplus}_R \mathfrak{g}_{6,8}$

- Describing representation: $R = ad\mathfrak{so}(3) \oplus 3D_0$

- Structure tensor:

$$
\begin{array}{llllll}
C_{12}^3 = 1, & C_{13}^2 = -1, & C_{23}^1 = 1, & C_{15}^6 = 1, & C_{16}^5 = -1, & C_{24}^6 = -1, \\
C_{26}^4 = 1, & C_{34}^5 = 1, & C_{35}^4 = 1, & C_{49}^4 = p, & C_{59}^5 = p, & C_{69}^6 = p, \\
C_{79}^7 = q, & C_{79}^8 = -1, & C_{89}^7 = 1, & C_{89}^8 = q.
\end{array}
$$

- $\mathrm{codim}_{\mathfrak{g}}\,[\mathfrak{g},\mathfrak{g}] = 1$.

- $\dim Der\,(\mathfrak{g}) = 11$.

- $\dim O\,(\mathfrak{g}) = 70$.

- Rank $A\,(\mathfrak{g}) = 6$

- Conditions on invariants:

$$\frac{\partial F}{\partial x_9} = 0.$$

- Codimension one subalgebra: $L\,\langle X_1,..,X_8\rangle \simeq L_{6,1} \oplus 2L_1$

- Invariants of subalgebra:

$$J_1 = x_4^2 + x_5^2 + x_6^2$$
$$J_2 = x_1 x_4 + x_2 x_5 + x_3 x_6$$
$$J_3 = x_7, \; J_4 = x_8$$

- Semi-invariance conditions:

$$\widehat{X}_9\,(J_1) = -2pJ_1, \; \widehat{X}_9\,(J_2) = -pJ_2$$
$$\widehat{X}_9\,(J_3) = -qJ_3 + J_4, \; \widehat{X}_9\,(J_4) = -J_3 - qJ_4.$$

- Invariants of $L_{9,3}^{p,q}$

$$I_1 = \frac{J_1}{J_2^2}, \; I_2 = \frac{J_2^{q-i}}{(J_3 - iJ_4)^p}, \; I_3 = \ln\left(x_7^2 + x_8^2\right) - 2q\arctan\left(\frac{J_4}{J_3}\right).$$

$\mathbf{L_{9,4}}$

- Levi decomposition: $L_{9,4} = \mathfrak{so}\,(3)\,\overrightarrow{\oplus}_R A_{6,5}$

- Describing representation: $R = R_4 \oplus 2D_0$

- Structure tensor:

$$C^3_{12} = 1, \quad C^2_{13} = -1, \quad C^1_{23} = 1, \quad C^7_{14} = \tfrac{1}{2}, \quad C^6_{15} = \tfrac{1}{2}, \quad C^5_{16} = -\tfrac{1}{2},$$
$$C^4_{17} = -\tfrac{1}{2}, \quad C^5_{24} = \tfrac{1}{2}, \quad C^4_{25} = -\tfrac{1}{2}, \quad C^7_{26} = \tfrac{1}{2}, \quad C^6_{27} = -\tfrac{1}{2}, \quad C^6_{34} = \tfrac{1}{2},$$
$$C^7_{35} = -\tfrac{1}{2}, \quad C^4_{36} = -\tfrac{1}{2}, \quad C^5_{37} = \tfrac{1}{2}, \quad C^9_{45} = -1, \quad C^8_{47} = 1, \quad C^8_{56} = -1,$$
$$C^9_{67} = 1.$$

- $\operatorname{codim}_{\mathfrak{g}}[\mathfrak{g},\mathfrak{g}] = 0$.

- $\dim Der(\mathfrak{g}) = 9$.

- $\dim O(\mathfrak{g}) = 72$.

- Rank $A(\mathfrak{g}) = 6$.

- Invariants of $L_{9,4}$

$$I_1 = x_8, \; I_2 = x_9, \; I_3 = \sqrt{D},$$

where D is the determinant

$$D := \begin{vmatrix} 0 & x_3 & -x_2 & \frac{x_7}{2} & \frac{x_6}{2} & -\frac{x_5}{2} & -\frac{x_4}{2} & x_1 \\ -x_3 & 0 & x_1 & \frac{x_5}{2} & -\frac{x_4}{2} & \frac{x_7}{2} & -\frac{x_6}{2} & x_2 \\ x_2 & -x_1 & 0 & \frac{x_6}{2} & -\frac{x_7}{2} & -\frac{x_4}{2} & \frac{x_5}{2} & x_3 \\ -\frac{x_7}{2} & -\frac{x_5}{2} & -\frac{x_6}{2} & 0 & -x_9 & 0 & x_8 & \frac{x_4}{2} \\ -\frac{x_4}{2} & \frac{x_4}{2} & \frac{x_7}{2} & x_9 & 0 & -x_8 & 0 & \frac{x_5}{2} \\ \frac{x_5}{2} & -\frac{x_7}{2} & \frac{x_4}{2} & 0 & x_8 & 0 & x_9 & \frac{x_6}{2} \\ \frac{x_4}{2} & \frac{x_6}{2} & -\frac{x_5}{2} & 0 & 0 & -x_9 & 0 & \frac{x_7}{2} \\ -x_1 & -x_2 & -x_3 & -\frac{x_4}{2} & -\frac{x_5}{2} & -\frac{x_6}{2} & -\frac{x_7}{2} & 0 \end{vmatrix}$$

$\mathbf{L}^p_{9,5}$ $[p \neq 0]$

- Levi decomposition: $L^p_{9,5} = \mathfrak{so}(3) \overrightarrow{\oplus}_R \mathfrak{g}_{6,1}$

- Describing representation: $R = R_4 \oplus 2D_0$

- Structure tensor:

$$C^3_{12} = 1, \quad C^2_{13} = -1, \quad C^1_{23} = 1, \quad C^7_{14} = \tfrac{1}{2}, \quad C^6_{15} = \tfrac{1}{2}, \quad C^5_{16} = -\tfrac{1}{2},$$
$$C^4_{17} = -\tfrac{1}{2}, \quad C^5_{24} = \tfrac{1}{2}, \quad C^4_{25} = -\tfrac{1}{2}, \quad C^7_{26} = \tfrac{1}{2}, \quad C^6_{27} = -\tfrac{1}{2}, \quad C^6_{34} = \tfrac{1}{2},$$
$$C^7_{35} = -\tfrac{1}{2}, \quad C^4_{36} = -\tfrac{1}{2}, \quad C^5_{37} = \tfrac{1}{2}, \quad C^4_{49} = 1, \quad C^5_{59} = 1, \quad C^6_{69} = 1,$$
$$C^7_{79} = 1, \quad C^8_{89} = p.$$

- $\operatorname{codim}_{\mathfrak{g}}[\mathfrak{g},\mathfrak{g}] = 1$.

- $\dim Der(\mathfrak{g}) = 13$.

- $\dim O(\mathfrak{g}) = 68$.

- Rank $A(\mathfrak{g}) = 8$

- Conditions on invariants:

$$\frac{\partial F}{\partial x_j} = 0, \ j = 1,2,3,9.$$

- Codimension one subalgebra: $L\langle X_1,..,X_8\rangle \simeq L_{7,2}\oplus L_1$

- Invariants of subalgebra:

$$J_1 = x_4^2 + x_5^2 + x_6^2 + x_7^2$$
$$J_2 = x_8$$

- Semi-invariance conditions:

$$\widehat{X}_9(J_1) = -2J_1, \ \widehat{X}_9(J_2) = -pJ_2.$$

- Invariants of $L_{9,5}^{p}$

$$I_1 = \frac{J_1^p}{J_2^2}.$$

$\mathbf{L}_{9,6}^{p,q}\ [q \neq 0]$

- Levi decomposition: $L_{9,6}^{p,q} = \mathfrak{so}(3)\overrightarrow{\oplus}_R \mathfrak{g}_{6,11}$

- Describing representation: $R = R_4 \oplus 2D_0$

- Structure tensor:

$$\begin{array}{llllll}
C_{12}^3 = 1, & C_{13}^2 = -1, & C_{23}^1 = 1, & C_{14}^7 = \tfrac{1}{2}, & C_{15}^6 = \tfrac{1}{2}, & C_{16}^5 = -\tfrac{1}{2}, \\
C_{17}^4 = -\tfrac{1}{2}, & C_{24}^5 = \tfrac{1}{2}, & C_{25}^4 = -\tfrac{1}{2}, & C_{26}^7 = \tfrac{1}{2}, & C_{27}^6 = -\tfrac{1}{2}, & C_{34}^6 = \tfrac{1}{2}, \\
C_{35}^7 = -\tfrac{1}{2}, & C_{36}^4 = -\tfrac{1}{2}, & C_{37}^5 = \tfrac{1}{2}, & C_{49}^4 = p, & C_{49}^6 = -1, & C_{59}^5 = p, \\
C_{59}^7 = -1, & C_{69}^6 = p, & C_{69}^4 = 1, & C_{79}^5 = 1, & C_{79}^7 = p, & C_{89}^8 = q.
\end{array}$$

- $\mathrm{codim}_{\mathfrak{g}}\,[\mathfrak{g},\mathfrak{g}] = 1$.

- $\dim Der\,(\mathfrak{g}) = 11$.

- $\dim O\,(\mathfrak{g}) = 70$.

- Rank $A\,(\mathfrak{g}) = 8$

- Conditions on invariants:

$$\frac{\partial F}{\partial x_j} = 0, \ j = 1,2,3,9.$$

- Codimension one subalgebra: $L\,\langle X_1,..,X_8\rangle \simeq L_{7,2} \oplus L_1$

- Invariants of subalgebra:

$$J_1 = x_4^2 + x_5^2 + x_6^2 + x_7^2$$
$$J_2 = x_8$$

- Semi-invariance conditions:

$$\widehat{X}_9\,(J_1) = -2pJ_1, \ \widehat{X}_9\,(J_2) = -qJ_2.$$

- Invariants of $L_{9,6}^{p;q}$

$$I_1 = \frac{J_1^q}{J_2^{2p}} \text{ for } p \neq 0,$$
$$I_1 = J_1 \text{ for } p = 0.$$

$\mathbf{L}_{9,7}^{p}\ (p \neq 0)$

- Levi decomposition: $L_{9,7}^{p} = \mathfrak{so}\,(3) \overset{\rightarrow}{\oplus}_R \mathfrak{g}_{6,82}^{*}$

- Describing representation: $R = R_4 \oplus 2D_0$

- Structure tensor:

$$
\begin{array}{llllll}
C^3_{12} = 1, & C^2_{13} = -1, & C^1_{23} = 1, & C^7_{14} = \tfrac{1}{2}, & C^6_{15} = \tfrac{1}{2}, & C^5_{16} = -\tfrac{1}{2}, \\
C^4_{17} = -\tfrac{1}{2}, & C^5_{24} = \tfrac{1}{2}, & C^4_{25} = -\tfrac{1}{2}, & C^7_{26} = \tfrac{1}{2}, & C^6_{27} = -\tfrac{1}{2}, & C^6_{34} = \tfrac{1}{2}, \\
C^7_{35} = -\tfrac{1}{2}, & C^4_{36} = -\tfrac{1}{2}, & C^5_{37} = \tfrac{1}{2}, & C^8_{46} = 1, & C^4_{49} = p, & C^6_{49} = 1, \\
C^8_{57} = 1, & C^5_{59} = p, & C^7_{59} = 1, & C^4_{69} = -1, & C^6_{69} = p, & C^5_{79} = -1, \\
C^7_{79} = p, & C^8_{89} = 2p.
\end{array}
$$

- $\operatorname{codim}_{\mathfrak{g}}\left[\mathfrak{g},\mathfrak{g}\right] = 1.$

- $\dim Der\left(\mathfrak{g}\right) = 10.$

- $\dim O\left(\mathfrak{g}\right) = 71.$

- Rank $A\left(\mathfrak{g}\right) = 8$

- Conditions on invariants:
$$
\frac{\partial F}{\partial x_9} = 0.
$$

- Codimension one subalgebra: $L\left\langle X_1,..,X_8\right\rangle \simeq L_{8,2}$

- Invariants of subalgebra:
$$
J_1 = x_8,
$$
$$
J_2 = \sqrt{D},
$$

where

$$
D = \begin{vmatrix}
0 & x_3 & -x_2 & \frac{x_7}{2} & \frac{x_6}{2} & -\frac{x_5}{2} & -\frac{x_4}{2} & x_1 \\
-x_3 & 0 & x_1 & \frac{x_5}{2} & -\frac{x_4}{2} & \frac{x_7}{2} & -\frac{x_6}{2} & x_2 \\
x_2 & -x_1 & 0 & \frac{x_6}{2} & -\frac{x_7}{2} & -\frac{x_4}{2} & \frac{x_5}{2} & x_3 \\
-\frac{x_7}{2} & -\frac{x_5}{2} & -\frac{x_6}{2} & 0 & 0 & x_8 & 0 & \frac{x_4}{2} \\
-\frac{x_4}{2} & \frac{x_4}{2} & \frac{x_7}{2} & 0 & 0 & 0 & x_8 & \frac{x_5}{2} \\
\frac{x_5}{2} & -\frac{x_7}{2} & \frac{x_4}{2} & -x_8 & 0 & 0 & 0 & \frac{x_6}{2} \\
\frac{x_4}{2} & \frac{x_6}{2} & -\frac{x_5}{2} & 0 & -x_8 & 0 & 0 & \frac{x_7}{2} \\
-x_1 & -x_2 & -x_3 & -\frac{x_4}{2} & -\frac{x_5}{2} & -\frac{x_6}{2} & -\frac{x_7}{2} & 0
\end{vmatrix}.
$$

- Semi-invariance conditions:

$$\widehat{X}_9\left(J_1\right)=-2pJ_1,\ \widehat{X}_9\left(J_2\right)=-4pJ_2.$$

- Invariants of $L_{9,7}^{p}$

$$I_1=\frac{J_2}{J_1^2}.$$

$\mathbf{L}_{9,7}^{0}$

- Levi decomposition: $L_{9,7}^0=\mathfrak{so}\left(3\right)\overrightarrow{\oplus}_R\mathfrak{g}_{6,82}^*$

- Describing representation: $R=R_4\oplus 2D_0$

- Structure tensor:

$$
\begin{aligned}
&C_{12}^3=1,\quad &&C_{13}^2=-1,\quad &&C_{23}^1=1,\quad &&C_{14}^7=\tfrac{1}{2},\quad &&C_{15}^6=\tfrac{1}{2},\quad &&C_{16}^5=-\tfrac{1}{2},\\
&C_{17}^4=-\tfrac{1}{2},\quad &&C_{24}^5=\tfrac{1}{2},\quad &&C_{25}^4=-\tfrac{1}{2},\quad &&C_{26}^7=\tfrac{1}{2},\quad &&C_{27}^6=-\tfrac{1}{2},\quad &&C_{34}^6=\tfrac{1}{2},\\
&C_{35}^7=-\tfrac{1}{2},\quad &&C_{36}^4=-\tfrac{1}{2},\quad &&C_{37}^5=\tfrac{1}{2},\quad &&C_{46}^8=1,\quad &&C_{49}^6=1,\quad &&C_{57}^8=1,\\
&C_{59}^7=1,\quad &&C_{69}^4=-1,\quad &&C_{79}^5=-1.
\end{aligned}
$$

- $\operatorname{codim}_{\mathfrak{g}}[\mathfrak{g},\mathfrak{g}]=1$.

- $\dim Der\left(\mathfrak{g}\right)=10$.

- $\dim O\left(\mathfrak{g}\right)=71$.

- Rank $A\left(\mathfrak{g}\right)=6$

- Codimension one subalgebra: $L\left\langle X_1,..,X_8\right\rangle\simeq L_{8,2}$

- Invariants of subalgebra:

$$
\begin{aligned}
J_1&=x_8,\\
J_2&=\sqrt{D},
\end{aligned}
$$

where

$$D = \begin{vmatrix} 0 & x_3 & -x_2 & \frac{x_7}{2} & \frac{x_6}{2} & -\frac{x_5}{2} & -\frac{x_4}{2} & x_1 \\ -x_3 & 0 & x_1 & \frac{x_5}{2} & -\frac{x_4}{2} & \frac{x_7}{2} & -\frac{x_6}{2} & x_2 \\ x_2 & -x_1 & 0 & \frac{x_6}{2} & -\frac{x_7}{2} & -\frac{x_4}{2} & \frac{x_5}{2} & x_3 \\ -\frac{x_7}{2} & -\frac{x_5}{2} & -\frac{x_6}{2} & 0 & 0 & x_8 & 0 & \frac{x_4}{2} \\ -\frac{x_4}{2} & \frac{x_4}{2} & \frac{x_7}{2} & 0 & 0 & 0 & x_8 & \frac{x_5}{2} \\ \frac{x_5}{2} & -\frac{x_7}{2} & \frac{x_4}{2} & -x_8 & 0 & 0 & 0 & \frac{x_6}{2} \\ \frac{x_4}{2} & \frac{x_6}{2} & -\frac{x_5}{2} & 0 & -x_8 & 0 & 0 & \frac{x_7}{2} \\ -x_1 & -x_2 & -x_3 & -\frac{x_4}{2} & -\frac{x_5}{2} & -\frac{x_6}{2} & -\frac{x_7}{2} & 0 \end{vmatrix}$$

- Semi-invariance conditions:

$$\widehat{X}_9\,(J_1) = 0,\ \widehat{X}_9\,(J_2) = 0.$$

- Invariants of $L_{9,7}^0$

$$I_1 = J_1,\ I_2 = J_2,$$
$$I_3 = x_4^5 + x_5^2 + x_6^2 + x_7^2 - 2x_8 x_9.$$

$L_{9,7}^*$

- Levi decomposition: $L_{9,7}^* = \mathfrak{so}\,(3) \overrightarrow{\oplus}_R \mathfrak{g}_{6,82}^{2,0,0}$

- Describing representation: $R = R_4 \oplus 2D_0$

- Structure tensor:

$$\begin{array}{llllll} C_{12}^3 = 1, & C_{13}^2 = -1, & C_{23}^1 = 1, & C_{14}^7 = \tfrac{1}{2}, & C_{15}^6 = \tfrac{1}{2}, & C_{16}^5 = -\tfrac{1}{2}, \\ C_{17}^4 = -\tfrac{1}{2}, & C_{24}^5 = \tfrac{1}{2}, & C_{25}^4 = -\tfrac{1}{2}, & C_{26}^7 = \tfrac{1}{2}, & C_{27}^6 = -\tfrac{1}{2}, & C_{34}^6 = \tfrac{1}{2}, \\ C_{35}^7 = -\tfrac{1}{2}, & C_{36}^4 = -\tfrac{1}{2}, & C_{37}^5 = \tfrac{1}{2}, & C_{46}^8 = 1, & C_{57}^8 = 1, & C_{49}^4 = 1, \\ C_{59}^5 = 1, & C_{69}^6 = 1, & C_{79}^7 = 1, & C_{89}^8 = 2. \end{array}$$

- $\mathrm{codim}_{\mathfrak{g}}\,[\mathfrak{g}, \mathfrak{g}] = 1.$

- $\dim Der\,(\mathfrak{g}) = 10.$

- $\dim O\,(\mathfrak{g}) = 71.$

- Rank $A\left(\mathfrak{g}\right) = 8$

- Conditions on invariants:

$$\frac{\partial F}{\partial x_9} = 0.$$

- Codimension one subalgebra: $L\left\langle X_1,..,X_8\right\rangle \simeq L_{8,2}$

- Invariants of subalgebra:

$$J_1 = x_8$$
$$J_2 = \sqrt{D},$$

where

$$D = \begin{vmatrix} 0 & x_3 & -x_2 & \frac{x_7}{2} & \frac{x_6}{2} & -\frac{x_5}{2} & -\frac{x_4}{2} & x_1 \\ -x_3 & 0 & x_1 & \frac{x_5}{2} & -\frac{x_4}{2} & \frac{x_7}{2} & -\frac{x_6}{2} & x_2 \\ x_2 & -x_1 & 0 & \frac{x_6}{2} & -\frac{x_7}{2} & -\frac{x_4}{2} & \frac{x_5}{2} & x_3 \\ -\frac{x_7}{2} & -\frac{x_5}{2} & -\frac{x_6}{2} & 0 & 0 & x_8 & 0 & \frac{x_4}{2} \\ -\frac{x_4}{2} & \frac{x_4}{2} & \frac{x_7}{2} & 0 & 0 & 0 & x_8 & \frac{x_5}{2} \\ \frac{x_5}{2} & -\frac{x_7}{2} & \frac{x_4}{2} & -x_8 & 0 & 0 & 0 & \frac{x_6}{2} \\ \frac{x_4}{2} & \frac{x_6}{2} & -\frac{x_5}{2} & 0 & -x_8 & 0 & 0 & \frac{x_7}{2} \\ -x_1 & -x_2 & -x_3 & -\frac{x_4}{2} & -\frac{x_5}{2} & -\frac{x_6}{2} & -\frac{x_7}{2} & 0 \end{vmatrix}$$

- Semi-invariance conditions:

$$\widehat{X}_9\left(J_1\right) = -2J_1,\ \widehat{X}_9\left(J_2\right) = -4J_2.$$

- Invariants of $L_{9,7}^{*}$

$$I_1 = \frac{J_2}{J_1^2}.$$

L$_{9,8}$

- Levi decomposition: $L_{9,8} = \mathfrak{so}\left(3\right) \overrightarrow{\oplus}_R \mathcal{N}_{6,18}^{0,1,1}$

- Describing representation: $R = R_4 \oplus 2D_0$

- Structure tensor:

$$C_{12}^3 = 1, \quad C_{13}^2 = -1, \quad C_{23}^1 = 1, \quad C_{14}^7 = \tfrac{1}{2}, \quad C_{15}^6 = \tfrac{1}{2}, \quad C_{16}^5 = -\tfrac{1}{2},$$
$$C_{17}^4 = -\tfrac{1}{2}, \quad C_{24}^5 = \tfrac{1}{2}, \quad C_{25}^4 = -\tfrac{1}{2}, \quad C_{26}^7 = \tfrac{1}{2}, \quad C_{27}^6 = -\tfrac{1}{2}, \quad C_{34}^6 = \tfrac{1}{2},$$
$$C_{35}^7 = -\tfrac{1}{2}, \quad C_{36}^4 = -\tfrac{1}{2}, \quad C_{37}^5 = \tfrac{1}{2}, \quad C_{48}^4 = 1, \quad C_{49}^6 = 1, \quad C_{58}^5 = 1,$$
$$C_{59}^7 = 1, \quad C_{68}^6 = 1, \quad C_{69}^4 = -1, \quad C_{78}^7 = 1, \quad C_{79}^5 = -1.$$

- $\operatorname{codim}_{\mathfrak{g}} [\mathfrak{g}, \mathfrak{g}] = 2.$

- $\dim Der(\mathfrak{g}) = 9.$

- $\dim O(\mathfrak{g}) = 72.$

- Rank $A(\mathfrak{g}) = 8$

- Condition on invariants.

$$\frac{\partial F}{\partial x_8} = 0.$$

- Codimension one subalgebra: $L\langle X_1, .., X_8\rangle \simeq L_{8,4}^0$

- Invariants of subalgebra:

$$J_1 = x_6^2 x_9 + 2x_3 x_6^2 + 4x_2 x_5 x_6 + 4x_1 x_6 x_7 - 4x_2 x_4 x_7 + x_5^2 x_9$$
$$\quad + 2x_3 x_4^2 + x_4^2 x_9 + 4x_1 x_4 x_5 + x_7^2 x_9 - 2x_3 x_5^2 - 2x_3 x_7^2$$
$$J_2 = x_4^2 + x_5^2 + x_6^2 + x_7^2$$

- Semi-invariance conditions:

$$\widehat{X}_9(J_1) = -2J_1, \ \widehat{X}_9(J_2) = -2J_2.$$

- Invariants of $L_{9,8}$

$$I_1 = \frac{J_1}{J_2}.$$

L$_{9,9}$

- Levi decomposition: $L_{9,9} = \mathfrak{so}(3) \overrightarrow{\oplus}_R \mathfrak{g}_{6,1}$

- Describing representation: $R = R_5 \oplus D_0$

- Structure tensor:

$$
\begin{aligned}
&C_{12}^3 = 1, && C_{13}^2 = -1, && C_{23}^1 = 1, && C_{14}^7 = \tfrac{1}{2}, && C_{15}^6 = -\tfrac{1}{2}, && C_{16}^5 = 2, \\
&C_{16}^8 = -1, && C_{17}^4 = -2, && C_{18}^6 = 3, && C_{24}^6 = \tfrac{1}{2}, && C_{25}^7 = \tfrac{1}{2}, && C_{26}^4 = -2, \\
&C_{27}^5 = -2, && C_{27}^8 = -1, && C_{28}^7 = 3, && C_{34}^{58} = 2, && C_{35}^4 = -2, && C_{36}^7 = 1, \\
&C_{37}^6 = -1, && C_{49}^4 = 1, && C_{59}^5 = 1, && C_{69}^6 = 1, && C_{79}^7 = 1, && C_{89}^8 = 1.
\end{aligned}
$$

- $\operatorname{codim}_{\mathfrak{g}}[\mathfrak{g},\mathfrak{g}] = 1$.

- $\dim Der(\mathfrak{g}) = 9$.

- $\dim O(\mathfrak{g}) = 72$.

- Rank $A(\mathfrak{g}) = 8$

- Conditions on invariants:

$$
\frac{\partial F}{\partial x_j} = 0, \ j = 1,2,3,9.
$$

- Codimension one subalgebra: $L\langle X_1,..,X_8\rangle \simeq L_{8,5}$

- Invariants of subalgebra:

$$
J_1 = 12\left(x_4^2 + x_5^2\right) + 3\left(x_6^2 + x_7^2\right) + x_8^2
$$
$$
J_2 = \frac{2}{9}x_8^3 + x_6^2 x_8 - 12x_4 x_6 x_7 - 8x_4^2 x_8 + 6x_5 x_6^2
$$
$$
- 6x_5 x_7^2 + x_7^2 x_8 - x_5^2 x_8.
$$

- Semi-invariance conditions:

$$
\widehat{X}_9(J_1) = -2J_1, \ \widehat{X}_9(J_2) = -3J_2.
$$

- Invariants of $L_{9,9}$

$$
I_1 = \frac{J_1^3}{J_2^2}.
$$

$L_{9,10}$

- Levi decomposition: $L_{9,10} = \mathfrak{so}\,(3) \overrightarrow{\oplus}_R 6L_1$

- Describing representation: $R = 2ad\mathfrak{so}\,(3)$

- Structure tensor:

$$
\begin{aligned}
&C_{12}^3 = 1, \quad C_{13}^2 = -1, \quad C_{23}^1 = 1, \quad C_{15}^6 = 1, \quad C_{16}^5 = -1, \quad C_{18}^9 = 1, \\
&C_{19}^8 - 1, \quad C_{24}^6 = -1, \quad C_{26}^4 = 1, \quad C_{27}^9 = -1, \quad C_{29}^7 = 1, \quad C_{34}^5 = 1, \\
&C_{35}^4 = -1, \quad C_{37}^8 = 1, \quad C_{38}^7 = -1.
\end{aligned}
$$

- $\operatorname{codim}_{\mathfrak{g}}[\mathfrak{g},\mathfrak{g}] = 0$.

- $\dim Der\,(\mathfrak{g}) = 13$.

- $\dim O\,(\mathfrak{g}) = 68$.

- Rank $A\,(\mathfrak{g}) = 6$

- Conditions on invariants:

$$
\frac{\partial F}{\partial x_j} = 0, \; j = 1,2,3.
$$

- Conditions on invariants:

$$
\frac{\partial F}{\partial x_i} = 0, \; i = 1,2,3
$$

- Invariants of $L_{9,10}$

$$
\begin{aligned}
I_1 &= x_4^2 + x_5^2 + x_6^2, \\
I_2 &= x_7^2 + x_8^2 + x_9^2, \\
I_3 &= x_4 x_7 + x_5 x_8 + x_6 x_9.
\end{aligned}
$$

$L_{9,11}$

- Levi decomposition: $L_{9,11} = \mathfrak{so}\,(3) \overrightarrow{\oplus}_R \mathcal{A}_{6,3}$

- Describing representation: $R = 2ad\mathfrak{so}\,(3)$

- Structure tensor:

$$
\begin{array}{llllll}
C_{12}^3 = 1, & C_{13}^2 = -1, & C_{23}^1 = 1, & C_{15}^6 = 1, & C_{16}^5 = -1, & C_{18}^9 = 1, \\
C_{19}^8 - 1, & C_{24}^6 = -1, & C_{26}^4 = 1, & C_{27}^9 = -1, & C_{29}^7 = 1, & C_{34}^5 = 1, \\
C_{35}^4 = -1, & C_{37}^8 = 1, & C_{38}^7 = -1, & C_{45}^9 = 1, & C_{46}^8 = -1, & C_{56}^7 = 1.
\end{array}
$$

- $\operatorname{codim}_{\mathfrak{g}}[\mathfrak{g},\mathfrak{g}] = 0.$

- $\dim Der\,(\mathfrak{g}) = 11.$

- $\dim O\,(\mathfrak{g}) = 70.$

- Rank $A\,(\mathfrak{g}) = 6$

- Invariants of $L_{9,11}$

$$
\begin{aligned}
I_1 &= x_7^2 + x_8^2 + x_9^2, \\
I_2 &= x_4 x_7 + x_5 x_8 + x_6 x_9, \\
I_3 &= x_1 x_7 + x_2 x_8 + x_3 x_9 + \frac{1}{2}\left(x_4^2 + x_5^2 + x_6^2\right).
\end{aligned}
$$

$\mathbf{L}_{9,12}$

- Levi decomposition: $L_{9,12} = \mathfrak{sl}\,(2,\mathbb{R}) \overrightarrow{\oplus}_R \mathcal{A}_{6,12}$

- Describing representation: $R = D_{\frac{1}{2}} \oplus 4D_0$

- Structure tensor:

$$
\begin{array}{llllll}
C_{12}^2 = 2, & C_{13}^3 = -2, & C_{23}^1 = 1, & C_{14}^4 = 1, & C_{15}^5 = -1, & C_{25}^4 = 1, \\
C_{34}^5 = 1, & C_{45}^6 = 1, & C_{79}^6 = 1, & C_{89}^7 = 1.
\end{array}
$$

- $\operatorname{codim}_{\mathfrak{g}}[\mathfrak{g},\mathfrak{g}] = 2.$

- $\dim Der\,(\mathfrak{g}) = 12.$

- $\dim O\,(\mathfrak{g}) = 69.$

- Rank $A(\mathfrak{g}) = 6$

- Conditions on invariants:

$$\frac{\partial F}{\partial x_9} = 0,$$

- Codimension one subalgebra:

$$L\langle X_1, .., X_8\rangle \simeq L_{6,2} \oplus 2L_1.$$

- Invariants of subalgebra:

$$J_1 = 4x_2x_3x_6 + 2x_2x_5^2 + 2x_1x_4x_5 - 2x_3x_4^2 + x_1x_6^2,$$
$$J_2 = x_6,$$
$$J_3 = x_7,$$
$$J_4 = x_8.$$

- Semi-invariance conditions:

$$\widehat{X}_9(J_1) = 0, \quad \widehat{X}_9(J_2) = 0,$$
$$\widehat{X}_9(J_3) = -J_2, \quad \widehat{X}_9(J_4) = -J_3.$$

- Invariants of $L_{9,12}$

$$I_1 = 2x_6x_8 - x_7^2,$$
$$I_2 = 4x_2x_3x_6 + 2x_2x_5^2 + 2x_1x_4x_5 - 2x_3x_4^2 + x_1x_6^2$$
$$I_3 = x_6.$$

$\mathbf{L}_{9,13}^{p,q,r} \ [pqr \neq 0]$

- Levi decomposition: $L_{9,13}^{p,q,r} = \mathfrak{sl}(2,\mathbb{R}) \overrightarrow{\oplus}_R \mathfrak{g}_{6,1}$

- Describing representation: $R = D_{\frac{1}{2}} \oplus 4D_0$

- Structure tensor:

$$C_{12}^2 = 2, \quad C_{13}^3 = -2, \quad C_{23}^1 = 1, \quad C_{14}^4 = 1, \quad C_{15}^5 = -1, \quad C_{25}^4 = 1,$$
$$C_{34}^5 = 1, \quad C_{49}^4 = 1, \quad C_{59}^5 = 1, \quad C_{69}^6 = p, \quad C_{79}^7 = q, \quad C_{89}^8 = r.$$

- $\operatorname{codim}_{\mathfrak{g}}[\mathfrak{g},\mathfrak{g}] = 1$.

- $\dim Der\left(\mathfrak{g}\right) = \begin{cases} 12, & p \neq q, q \neq r, r \neq p \\ 14, & p = q \text{ or } p = r \text{ or } q = r \\ 18, & p = q = r \end{cases}$.

- $\dim O\left(\mathfrak{g}\right) = \begin{cases} 69, & p \neq q, q \neq r, r \neq p \\ 67, & p = q \text{ or } q = r \text{ or } r = p \\ 63, & p = q = r \end{cases}$.

- $\operatorname{Rank} A\left(\mathfrak{g}\right) = 6$

- Conditions on invariants:
$$\frac{\partial F}{\partial x_9} = 0,$$

- Codimension one subalgebra:
$$L\langle X_1, .., X_8 \rangle \simeq \mathfrak{sa}\left(2, \mathbb{R}\right) \oplus 3L_1$$

- Invariants of subalgebra:
$$J_1 = x_3 x_4^2 - x_1 x_4 x_5 - x_2 x_5^2,$$
$$J_2 = x_6,$$
$$J_3 = x_7,$$
$$J_4 = x_8.$$

- Semi-invariance conditions:
$$\widehat{X}_9\left(J_1\right) = -2J_1, \ \widehat{X}_9\left(J_2\right) = -pJ_2,$$
$$\widehat{X}_9\left(J_3\right) = -qJ_3, \ \widehat{X}_9\left(J_4\right) = -rJ_4.$$

- Invariants of $L_{9,13}^{p,q,r}$
$$I_1 = \frac{J_1^p}{J_2^2}, \ I_2 = \frac{J_1^q}{J_3^2}, \ I_3 = \frac{J_1^r}{J_4^2}.$$

$\mathbf{L}_{9,14}^{p,q}\ [p \neq 0]$

- Levi decomposition: $L_{9,14}^{p,q} = \mathfrak{sl}(2,\mathbb{R}) \overrightarrow{\oplus}_R \mathfrak{g}_{6,2}$

- Describing representation: $R = D_{\frac{1}{2}} \oplus 4D_0$

- Structure tensor:

$$C_{12}^2 = 2, \quad C_{13}^3 = -2, \quad C_{23}^1 = 1, \quad C_{14}^4 = 1, \quad C_{15}^5 = -1, \quad C_{25}^4 = 1,$$
$$C_{34}^5 = 1, \quad C_{49}^4 = 1, \quad C_{59}^5 = 1, \quad C_{69}^6 = p, \quad C_{79}^7 = q, \quad C_{89}^7 = 1,$$
$$C_{89}^8 = q.$$

- $\mathrm{codim}_{\mathfrak{g}}\,[\mathfrak{g},\mathfrak{g}] = \begin{cases} 1, & q \neq 0 \\ 2, & q = 0 \end{cases}.$

- $\dim Der\,(\mathfrak{g}) = \begin{cases} 12, & p \neq q \\ 14, & p = q \end{cases}.$

- $\dim O\,(\mathfrak{g}) = \begin{cases} 69, & p \neq q \\ 67, & p = q \end{cases}.$

- Rank $A\,(\mathfrak{g}) = 6$

- Conditions on invariants:

$$\frac{\partial F}{\partial x_9} = 0,$$

- Codimension one subalgebra:

$$L\langle X_1,..,X_8\rangle \simeq \mathfrak{sa}\,(2,\mathbb{R}) \oplus 3L_1$$

- Invariants of subalgebra:

$$J_1 = x_3 x_4^2 - x_1 x_4 x_5 - x_2 x_5^2,$$
$$J_2 = x_6,$$
$$J_3 = x_7,$$
$$J_4 = x_8.$$

- Semi-invariance conditions:

$$\widehat{X}_9\left(J_1\right) = -2J_1,\ \widehat{X}_9\left(J_2\right) = -pJ_2,$$
$$\widehat{X}_9\left(J_3\right) = -qJ_3,\ \widehat{X}_9\left(J_4\right) = -J_3 - qJ_4.$$

- Invariants of $L_{9,14}^{p,q}$

$$I_1 = \frac{J_1^p}{J_2^2},\ I_2 = \frac{J_1^q}{J_3^2},\ I_3 = \frac{J_4}{J_3} - \frac{1}{q}\ln J_3,\ (q \neq 0),$$

$$I_1 = \frac{J_1^p}{J_2^2},\ I_2 = J_3,\ I_3 = \ln\left(J_2 J_3\right) - \frac{J_4}{J_3},\ (q = 0).$$

$\mathbf{L}_{9,15}^p$

- Levi decomposition: $L_{9,15}^p = \mathfrak{sl}\left(2,\mathbb{R}\right) \overrightarrow{\oplus}_R \mathfrak{g}_{6,3}$

- Describing representation: $R = D_{\frac{1}{2}} \oplus 4D_0$

- Structure tensor:

$$
\begin{array}{llllll}
C_{12}^2 = 2, & C_{13}^3 = -2, & C_{23}^1 = 1, & C_{14}^4 = 1, & C_{15}^5 = -1, & C_{25}^4 = 1, \\
C_{34}^5 = 1, & C_{49}^4 = 1, & C_{59}^5 = 1, & C_{69}^6 = p, & C_{79}^6 = 1, & C_{79}^7 = p, \\
C_{89}^7 = 1, & C_{89}^8 = p.
\end{array}
$$

- $\mathrm{codim}_{\mathfrak{g}}\left[\mathfrak{g},\mathfrak{g}\right] = \begin{cases} 1, & p \neq 0 \\ 2, & p = 0 \end{cases}$.

- $\dim Der\left(\mathfrak{g}\right) = 12$.

- $\dim O\left(\mathfrak{g}\right) = 69$.

- Rank $A\left(\mathfrak{g}\right) = 6$

- Conditions on invariants:

$$\frac{\partial F}{\partial x_9} = 0,$$

- Codimension one subalgebra:

$$L \langle X_1, .., X_8 \rangle \simeq \mathfrak{sa}(2, \mathbb{R}) \oplus 3L_1$$

- Invariants of subalgebra:

$$J_1 = x_3 x_4^2 - x_1 x_4 x_5 - x_2 x_5^2,$$
$$J_2 = x_6,$$
$$J_3 = x_7,$$
$$J_4 = x_8.$$

- Semi-invariance conditions:

$$\widehat{X}_9(J_1) = -2J_1, \ \widehat{X}_9(J_2) = -pJ_2,$$
$$\widehat{X}_9(J_3) = -J_2 - pJ_3, \ \widehat{X}_9(J_4) = -J_3 - pJ_4.$$

- Invariants of $L_{9,15}^p$

$$I_1 = \frac{J_1^p}{J_2^2}, \ I_2 = \frac{2J_4}{J_2} - \left(\frac{J_3}{J_2}\right)^2, \ I_3 = 2\frac{J_4}{J_2} + \frac{1}{p^2}\ln^2 J_2, \ (p \neq 0),$$

$$I_1 = J_2, \ I_2 = \frac{2}{J_2 J_4} - \left(\frac{J_3}{J_2 J_4}\right)^2, \ I_3 = \frac{2J_3}{J_2} + \ln\left(\frac{J_2}{J_1}\right), \ (p = 0).$$

$\mathbf{L}_{9,16}^{p,q,r} \ [pq \neq 0, \ r \geq 0]$

- Levi decomposition: $L_{9,16}^{p,q,r} = \mathfrak{sl}(2, \mathbb{R}) \overrightarrow{\oplus}_R \mathfrak{g}_{6,8}$

- Describing representation: $R = D_{\frac{1}{2}} \oplus 4D_0$

- Structure tensor:

$$\begin{aligned}
&C_{12}^2 = 2, \ \ C_{13}^3 = -2, \ \ C_{23}^1 = 1, \ \ C_{14}^4 = 1, \ \ C_{15}^5 = -1, \ \ C_{25}^4 = 1, \\
&C_{34}^5 = 1, \ \ C_{49}^4 = p, \ \ \ \ C_{59}^5 = p, \ \ C_{69}^6 = q, \ \ C_{79}^7 = r, \ \ \ \ \ C_{79}^8 = -1, \\
&C_{89}^7 = 1, \ \ C_{89}^8 = r.
\end{aligned}$$

- $\mathrm{codim}_{\mathfrak{g}}[\mathfrak{g}, \mathfrak{g}] = 1.$

- $\dim Der\left(\mathfrak{g}\right)=12.$

- $\dim O\left(\mathfrak{g}\right)=69.$

- Rank $A\left(\mathfrak{g}\right)=6$

- Conditions on invariants:

$$\frac{\partial F}{\partial x_9}=0,$$

- Codimension one subalgebra:

$$L\langle X_1,..,X_8\rangle \simeq \mathfrak{sa}\left(2,\mathbb{R}\right)\oplus 3L_1$$

- Invariants of subalgebra:

$$J_1 = x_3 x_4^2 - x_1 x_4 x_5 - x_2 x_5^2,$$
$$J_2 = x_6,$$
$$J_3 = x_7,$$
$$J_4 = x_8.$$

- Semi-invariance conditions:

$$\widehat{X}_9\left(J_1\right)=-2pJ_1,\ \widehat{X}_9\left(J_2\right)=-qJ_2,$$
$$\widehat{X}_9\left(J_3\right)=-rJ_3+J_4,\ \widehat{X}_9\left(J_4\right)=-J_3-rJ_4.$$

- Invariants of $L_{9,16}^{p,q,r}$

$$I_1 = \frac{J_1^p}{J_2^2},\ I_2 = \left(J_3^2 + J_4^2\right)\left(\frac{J_3 + iJ_4}{J_3 - iJ_4}\right)^{ri},\ I_3 = J_2\exp\left(q\arctan\left(\frac{J_3}{J_4}\right)\right).$$

$\mathbf{L}_{9,17}^{p,q}\ [pq\neq 0]$

- Levi decomposition: $L_{9,17}^{p,q}=\mathfrak{sl}\left(2,\mathbb{R}\right)\overrightarrow{\oplus}_R\mathfrak{g}_{6,13}$

- Describing representation: $R=D_{\frac{1}{2}}\oplus 4D_0$

- Structure tensor:

$$C_{12}^2 = 2, \quad C_{13}^3 = -2, \quad C_{23}^1 = 1, \quad C_{14}^4 = 1, \quad C_{15}^5 = -1, \quad C_{25}^4 = 1,$$
$$C_{34}^5 = 1, \quad C_{45}^6 = 1, \quad C_{49}^4 = 1, \quad C_{59}^5 = 1, \quad C_{69}^6 = 2, \quad C_{79}^7 = p,$$
$$C_{89}^8 = q.$$

- $\operatorname{codim}_{\mathfrak{g}}[\mathfrak{g},\mathfrak{g}] = 1.$

- $\dim Der(\mathfrak{g}) = \begin{cases} 11, & p \neq q, p \neq 2, q \neq 2 \\ 12, & p \neq q,\ p = 2 \text{ or } q = 2 \\ 13, & p = q,\ p \neq 2 \\ 15, & p = q = 2 \end{cases}.$

- $\dim O(\mathfrak{g}) = \begin{cases} 70, & p \neq q,\ p \neq 2,\ q \neq 2 \\ 69, & p \neq q,\ p = 2 \text{ or } q = 2 \\ 68, & p = q,\ p \neq 2 \\ 66, & p = q = 2 \end{cases}.$

- Rank $A(\mathfrak{g}) = 6$

- Conditions on invariants:

$$\frac{\partial F}{\partial x_9} = 0,$$

- Codimension one subalgebra:

$$L\langle X_1,..,X_8\rangle \simeq L_{6,2} \oplus 2L_1.$$

- Invariants of subalgebras:

$$J_1 = 2x_2 x_3 x_6 + x_2 x_5^2 + x_1 x_4 x_5 - x_3 x_4^2 + \frac{1}{2}x_1^2 x_6,$$
$$J_2 = x_6, \quad J_3 = x_7, \quad J_4 = x_8.$$

- Semi-invariance conditions:

$$\widehat{X}_9(J_1) = -2J_1,\ \widehat{X}_9(J_2) = -2J_2,$$
$$\widehat{X}_9(J_3) = -pJ_3,\ \widehat{X}_9(J_4) = -qJ_4.$$

- Invariants of $L_{9,17}^{p,q}$

$$I_1 = \frac{J_1}{J_2}, \quad I_2 = \frac{J_3^2}{J_2^p}, \quad I_3 = \frac{J_4^2}{J_3^q}.$$

$L_{9,18}^{p,q}$ $[p^2 + q^2 \neq 0]$

- Levi decomposition: $L_{9,18}^{p,q} = \mathfrak{sl}(2,\mathbb{R}) \overrightarrow{\oplus}_R \mathfrak{g}_{6,13}$

- Describing representation: $R = D_{\frac{1}{2}} \oplus 4D_0$

- Structure tensor:

$$
\begin{aligned}
&C_{12}^2 = 2, \quad C_{13}^3 = -2, \quad C_{23}^1 = 1, \quad C_{14}^4 = 1, \qquad C_{15}^5 = -1, \quad C_{25}^4 = 1, \\
&C_{34}^5 = 1, \quad C_{49}^4 = 1, \qquad C_{59}^5 = 1, \quad C_{69}^6 = p+q, \quad C_{78}^6 = 1, \qquad C_{79}^7 = p, \\
&C_{89}^8 = q.
\end{aligned}
$$

- $\operatorname{codim}_{\mathfrak{g}}[\mathfrak{g},\mathfrak{g}] = \begin{cases} 1, & pq \neq 0 \\ 2, & p=0 \text{ or } q=0 \end{cases}.$

- $\dim Der(\mathfrak{g}) = \begin{cases} 11, & p \neq q \\ 13, & p = q \end{cases}.$

- $\dim O(\mathfrak{g}) = \begin{cases} 70, & p \neq q \\ 68, & p = q \end{cases}.$

- Rank $A(\mathfrak{g}) = 8$

- Conditions on invariants:

$$\frac{\partial F}{\partial x_i} = 0, \; i = 7,8,9.$$

- Codimension three subalgebra:

$$L\langle X_1,..,X_6 \rangle \simeq L_{5,1} \oplus L_1$$

- Invariants of subalgebra:

$$
\begin{aligned}
J_1 &= x_3 x_4^2 - x_1 x_4 x_5 - x_2 x_5^2, \\
J_2 &= x_6,
\end{aligned}
$$

- Semi-invariance conditions:

$$\widehat{X}_9(J_1) = -2J_1, \ \widehat{X}_9(J_2) = -(p+q)J_2.$$

- Invariants of $L_{9,18}^{p,q}$

$$I_1 = \frac{J_1^{p+q}}{J_2^2}, \ p+q \neq 0,$$

$$I_1 = J_2, \ p+q = 0.$$

$\mathbf{L}_{9,19}^{p}$

- Levi decomposition: $L_{9,19}^{p} = \mathfrak{sl}(2,\mathbb{R}) \overrightarrow{\oplus}_{R} \mathfrak{g}_{6,14}$

- Describing representation: $R = D_{\frac{1}{2}} \oplus 4D_0$

- Structure tensor:

$$\begin{aligned}
&C_{12}^2 = 2, \quad C_{13}^3 = -2, \quad C_{23}^1 = 1, \quad C_{14}^4 = 1, \quad C_{15}^5 = -1, \quad C_{25}^4 = 1, \\
&C_{34}^5 = 1, \quad C_{49}^4 = p, \quad C_{59}^5 = p, \quad C_{69}^6 = 2p, \quad C_{79}^7 = 1, \quad C_{89}^6 = 1, \\
&C_{89}^8 = 2p, \quad C_{45}^6 = 1.
\end{aligned}$$

- $\mathrm{codim}_{\mathfrak{g}}[\mathfrak{g},\mathfrak{g}] = \begin{cases} 1, & p \neq 0 \\ 2, & p = 0 \end{cases}$.

- $\dim Der(\mathfrak{g}) = \begin{cases} 11, & p \neq \frac{1}{2} \\ 13, & p = \frac{1}{2} \end{cases}$.

- $\dim O(\mathfrak{g}) = \begin{cases} 70, & p \neq \frac{1}{2} \\ 68, & p = \frac{1}{2} \end{cases}$.

- Rank $A(\mathfrak{g}) = 6$

- Conditions on invariants:

$$\frac{\partial F}{\partial x_9} = 0,$$

- Codimension one subalgebra:

$$L \langle X_1, .., X_8 \rangle \simeq L_{6,2} \oplus 2L_1$$

- Invariants of subalgebra:

$$J_1 = 2x_2 x_3 x_6 + x_2 x_5^2 + x_1 x_4 x_5 - x_3 x_4^2 + \frac{1}{2} x_1^2 x_6,$$

$$J_2 = x_6,$$

$$J_3 = x_7,$$

$$J_4 = x_8.$$

- Semi-invariance conditions:

$$\widehat{X}_9(J_1) = -2pJ_1, \ \widehat{X}_9(J_2) = -2pJ_2,$$

$$\widehat{X}_9(J_3) = -J_3, \ \widehat{X}_9(J_4) = -J_2 - pJ_4.$$

- Invariants of $L_{9,19}^p$

$$I_1 = \frac{J_1}{J_2}, \ I_2 = \frac{J_3^{2p}}{J_2}, \ I_3 = -2p\frac{J_4}{J_2} + \frac{1}{p}\ln J_2, \ (p \neq 0),$$

$$I_1 = J_1, \ I_2 = J_2, \ I_3 = J_4 - J_2 \ln J_3, \ (p = 0).$$

$\mathbf{L}_{9,20}^{p}$

- Levi decomposition: $L_{9,20}^p = \mathfrak{sl}(2, \mathbb{R}) \overrightarrow{\oplus}_R \mathfrak{g}_{6,21}$

- Describing representation: $R = D_{\frac{1}{2}} \oplus 4D_0$

- Structure tensor:

$$C_{12}^2 = 2, \ C_{13}^3 = -2, \ C_{23}^1 = 1, \ C_{14}^4 = 1, \ C_{15}^5 = -1, \ C_{25}^4 = 1,$$
$$C_{34}^5 = 1, \ C_{49}^4 = 1, \ C_{59}^5 = 1, \ C_{69}^6 = 2p, \ C_{78}^6 = 1, \ C_{79}^7 = p,$$
$$C_{89}^7 = 1, \ C_{89}^8 = p.$$

- $\operatorname{codim}_{\mathfrak{g}} [\mathfrak{g}, \mathfrak{g}] = \begin{cases} 1, & p \neq 0 \\ 2, & p = 0 \end{cases}.$

- $\dim Der(\mathfrak{g}) = 11$.

- $\dim O(\mathfrak{g}) = 70$.

- Rank $A(\mathfrak{g}) = 8$

- Conditions on invariants:

$$\frac{\partial F}{\partial x_i} = 0, \ i = 7,8,9.$$

- Codimension three subalgebra:

$$L\langle X_1,..,X_6\rangle \simeq L_{5,1} \oplus L_1$$

- Invariants of subalgebra:

$$J_1 = x_3 x_4^2 - x_1 x_4 x_5 - x_2 x_5^2,$$
$$J_2 = x_6.$$

- Semi-invariance conditions:

$$\widehat{X}_9(J_1) = -2J_1, \ \widehat{X}_9(J_2) = -2pJ_2.$$

- Invariants of $L_{9,20}^p$

$$I_1 = \frac{J_1^{2p}}{J_2^2}, \ (p \neq 0),$$
$$I_1 = J_2, \ (p = 0).$$

$\mathbf{L}_{9,21}^p$

- Levi decomposition: $L_{9,21}^p = \mathfrak{sl}(2,\mathbb{R}) \overrightarrow{\oplus}_R \mathfrak{g}_{6,25}$

- Describing representation: $R = D_{\frac{1}{2}} \oplus 4D_0$

- Structure tensor:

$$
\begin{aligned}
&C^2_{12} = 2, \quad C^3_{13} = -2, \quad C^1_{23} = 1, \quad C^4_{14} = 1, \quad C^5_{15} = -1, \quad C^4_{25} = 1, \\
&C^5_{34} = 1, \quad C^4_{49} = 1, \quad\;\; C^5_{59} = 1, \quad C^6_{69} = 2, \quad C^7_{79} = p, \quad\;\;\; C^7_{89} = 1, \\
&C^8_{89} = p, \quad C^6_{45} = 1.
\end{aligned}
$$

- $\operatorname{codim}_{\mathfrak{g}} [\mathfrak{g}, \mathfrak{g}] = \begin{cases} 1, & p \neq 0 \\ 2, & p = 0 \end{cases}$.

- $\dim Der(\mathfrak{g}) = \begin{cases} 11, & p \neq 2 \\ 12, & p = 2 \end{cases}$.

- $\dim O(\mathfrak{g}) = \begin{cases} 70, & p \neq 2 \\ 69, & p = 2 \end{cases}$.

- $\operatorname{Rank} A(\mathfrak{g}) = 6$

- Conditions on invariants:

$$
\frac{\partial F}{\partial x_9} = 0,
$$

- Codimension one subalgebra:

$$
L \langle X_1, .., X_8 \rangle \simeq L_{6,2} \oplus 2L_1
$$

- Invariants of subalgebra:

$$
\begin{aligned}
J_1 &= 2x_2 x_3 x_6 + x_2 x_5^2 + x_1 x_4 x_5 - x_3 x_4^2 + \frac{1}{2} x_1^2 x_6, \\
J_2 &= x_6, \\
J_3 &= x_7, \\
J_4 &= x_8.
\end{aligned}
$$

- Semi-invariance conditions:

$$
\begin{aligned}
&\widehat{X}_9 (J_1) = -2J_1, \; \widehat{X}_9 (J_2) = -2J_2, \\
&\widehat{X}_9 (J_3) = -pJ_3, \; \widehat{X}_9 (J_4) = -J_3 - pJ_4.
\end{aligned}
$$

- Invariants of $L_{9,21}^p$

$$I_1 = \frac{J_1}{J_2}, \; J_2 = \frac{J_3^2}{J_2^p}, \; J_3 = \frac{\left(-2J_4 + J_3 \ln\left(J_2\right)\right)^2}{J_2^p}.$$

L$_{9,22}$

- Levi decomposition: $L_{9,22} = \mathfrak{sl}\left(2, \mathbb{R}\right) \overrightarrow{\oplus}_R \mathfrak{g}_{6,26}$

- Describing representation: $R = D_{\frac{1}{2}} \oplus 4D_0$

- Structure tensor:

$$\begin{array}{llllll}
C_{12}^2 = 2, & C_{13}^3 = -2, & C_{23}^1 = 1, & C_{14}^4 = 1, & C_{15}^5 = -1, & C_{25}^4 = 1, \\
C_{34}^5 = 1, & C_{45}^6 = 1, & C_{49}^4 = 1, & C_{59}^5 = 1, & C_{69}^6 = 2, & C_{79}^6 = 1, \\
C_{79}^7 = 2, & C_{89}^7 = 1, & C_{89}^8 = 2.
\end{array}$$

- $\mathrm{codim}_{\mathfrak{g}}\left[\mathfrak{g}, \mathfrak{g}\right] = 1$.

- $\dim Der\left(\mathfrak{g}\right) = 11$.

- $\dim O\left(\mathfrak{g}\right) = 70$.

- Rank $A\left(\mathfrak{g}\right) = 6$

- Conditions on invariants:

$$\frac{\partial F}{\partial x_9} = 0,$$

- Codimension one subalgebra:

$$L\left\langle X_1, .., X_8 \right\rangle \simeq L_{6,2} \oplus 2L_1$$

- Invariants of subalgebra:

$$J_1 = 2x_2x_3x_6 + x_2x_5^2 + x_1x_4x_5 - x_3x_4^2 + \frac{1}{2}x_1^2x_6,$$

$$J_2 = x_6,$$

$$J_3 = x_7,$$

$$J_4 = x_8.$$

- Semi-invariance conditions:

$$\widehat{X}_9\left(J_1\right) = -2J_1, \ \widehat{X}_9\left(J_2\right) = -2J_2,$$
$$\widehat{X}_9\left(J_3\right) = -J_2 - 2J_3, \ \widehat{X}_9\left(J_4\right) = -J_3 - 2J_4.$$

- Invariants of $L_{9,22}$

$$I_1 = \frac{J_1}{J_2}, \ I_2 = \frac{-2J_3 + J_2\ln\left(J_2\right)}{J_2}, \ I_3 = \frac{8J_4 + J_2\ln\left(J_2\right)^2 - 4J_3\ln\left(J_2\right)}{J_2}.$$

$\mathbf{L}_{9,23}^{p,q}$ $\left[p \neq 0\right]$

- Levi decomposition: $L_{9,23}^{p,q} = \mathfrak{sl}\left(2,\mathbb{R}\right) \overrightarrow{\oplus}_R \mathfrak{g}_{6,32}$

- Describing representation: $R = D_{\frac{1}{2}} \oplus 4D_0$

- Structure tensor:

$$\begin{aligned}
&C_{12}^2 = 2, &&C_{13}^3 = -2, &&C_{23}^1 = 1, &&C_{14}^4 = 1, &&C_{15}^5 = -1, &&C_{25}^4 = 1,\\
&C_{34}^5 = 1, &&C_{78}^6 = 1, &&C_{49}^4 = p, &&C_{59}^5 = p, &&C_{69}^6 = 2q, &&C_{79}^7 = q,\\
&C_{79}^8 = -1, &&C_{89}^7 = 1, &&C_{89}^8 = q.
\end{aligned}$$

- $\mathrm{codim}_\mathfrak{g}\left[\mathfrak{g},\mathfrak{g}\right] = 1$.

- $\dim Der\left(\mathfrak{g}\right) = 11$.

- $\dim O\left(\mathfrak{g}\right) = 70$.

- Rank $A\left(\mathfrak{g}\right) = 8$

- Conditions on invariants:

$$\frac{\partial F}{\partial x_j} = 0, \ j = 7,8,9$$

- Codimension one subalgebra:

$$L\left\langle X_1,..,X_8\right\rangle \simeq L_{5,1} \oplus \mathfrak{h}_1$$

- Invariants of subalgebra:

$$J_1 = -x_2 x_5^2 - x_1 x_4 x_5 + x_3 x_4^2,$$
$$J_2 = x_6,$$

- Semi-invariance conditions:

$$\widehat{X}_9(J_1) = -pJ_1, \ \widehat{X}_9(J_2) = -qJ_2.$$

- Invariants of $L_{9,23}^{p,q}$

$$I_1 = \frac{J_1^q}{J_2^p}, \ q \neq 0$$
$$I_1 = J_2, \ q = 0.$$

$\mathbf{L}_{9,24}^{p,q}$ $[p \neq 0, \ q \geq 0]$

- Levi decomposition: $L_{9,24}^{p,q} = \mathfrak{sl}(2,\mathbb{R}) \overrightarrow{\oplus}_R \mathfrak{g}_{6,35}$

- Describing representation: $R = D_{\frac{1}{2}} \oplus 4D_0$

- Structure tensor:

$$\begin{array}{llllll}
C_{12}^2 = 2, & C_{13}^3 = -2, & C_{23}^1 = 1, & C_{14}^4 = 1, & C_{15}^5 = -1, & C_{25}^4 = 1, \\
C_{34}^5 = 1, & C_{45}^6 = 1, & C_{49}^4 = p, & C_{59}^5 = p, & C_{69}^6 = 2p, & C_{79}^7 = q, \\
C_{79}^8 = -1, & C_{89}^7 = 1, & C_{89}^8 = q. &&&
\end{array}$$

- $\operatorname{codim}_{\mathfrak{g}}[\mathfrak{g},\mathfrak{g}] = 1.$

- $\dim Der(\mathfrak{g}) = 11.$

- $\dim O(\mathfrak{g}) = 70.$

- Rank $A(\mathfrak{g}) = 6$

- Conditions on invariants:

$$\frac{\partial F}{\partial x_9} = 0,$$

- Codimension one subalgebra:

$$L\langle X_1,..,X_8\rangle \simeq L_{6,2}\oplus 2L_1$$

- Invariants of subalgebra:

$$J_1 = 2x_2x_3x_6 + x_2x_5^2 + x_1x_4x_5 - x_3x_4^2 + \frac{1}{2}x_1^2x_6,$$
$$J_2 = x_6,$$
$$J_3 = x_7,$$
$$J_4 = x_8.$$

- Semi-invariance conditions:

$$\widehat{X}_9(J_1) = -2pJ_1,\ \widehat{X}_9(J_2) = -2pJ_2,$$
$$\widehat{X}_9(J_3) = -qJ_3 - J_4,\ \widehat{X}_9(J_4) = -J_3 - qJ_4.$$

- Invariants of $L_{9,24}^{p,q}$

$$I_1 = \frac{J_1}{J_2},\ I_2 = \left(J_3^2 + J_4^2\right)\left(\frac{J_3 - iJ_4}{J_3 + iJ_4}\right)^{-iq},\ I_3 = J_2\exp\left(2p\arctan\left(J_3J_4^{-1}\right)\right).$$

$L_{9,25}^p\ [|p|\leq 1]$

- Levi decomposition: $L_{9,25}^p = \mathfrak{sl}(2,\mathbb{R})\overrightarrow{\oplus}_R \mathfrak{g}_{6,82}$

- Describing representation: $R = D_{\frac{1}{2}}\oplus 4D_0$

- Structure tensor:

$$\begin{array}{llllll}
C_{12}^2 = 2, & C_{13}^3 = -2, & C_{23}^1 = 1, & C_{14}^4 = 1, & C_{15}^5 = -1, & C_{25}^4 = 1,\\
C_{34}^5 = 1, & C_{45}^6 = 1, & C_{49}^4 = p, & C_{59}^5 = p, & C_{69}^6 = 2p, & C_{79}^7 = 1,\\
C_{89}^8 = 2p-1, & C_{78}^6 = 1. & & & &
\end{array}$$

- $\mathrm{codim}_{\mathfrak{g}}[\mathfrak{g},\mathfrak{g}] = \left\{\begin{array}{ll} 1, & p\neq\frac{1}{2}\\ 2, & p=\frac{1}{2}\end{array}\right.$.

- $\dim Der\,(\mathfrak{g}) = \begin{cases} 10, & p \neq 1 \\ 12, & p = 1 \end{cases}$.

- $\dim O\,(\mathfrak{g}) = \begin{cases} 71, & p \neq 1 \\ 69, & p = 1 \end{cases}$.

- $\operatorname{Rank} A\,(\mathfrak{g}) = \begin{cases} 8 & \text{if } p \neq 0 \\ 6 & \text{if } p = 0 \end{cases}$.

- Conditions on invariants:

$$\frac{\partial F}{\partial x_j} = 0, \ j = 7,8,9 \text{ if } p \neq 0$$

- Codimension one subalgebra for $p \neq 0$:

$$L\,\langle X_1,..,X_8 \rangle \simeq L_{8,6}$$

- Invariants of subalgebra $(p \neq 0)$:

$$J_1 = 2x_2 x_3 x_8 + x_2 x_5^2 + x_1 x_4 x_5 - x_3 x_4^2 + \frac{1}{2} x_1^2 x_8,$$
$$J_2 = x_6.$$

- Semi-invariance conditions:

$$\widehat{X}_9\,(J_1) = -2pJ_1, \ \widehat{X}_9\,(J_2) = -2pJ_2.$$

- Invariants of $L^p_{9,25}$

$$I_1 = \frac{J_1}{J_2}, \ p \neq 0,$$
$$I_1 = J_1, \ I_2 = J_2, \ I_3 = x_6 x_9 - x_7 x_8, \ p = 0.$$

$\mathbf{L}^{\varepsilon}_{9,26} \ [\varepsilon = \pm 1]$

- Levi decomposition: $L^{\varepsilon}_{9,26} = \mathfrak{sl}\,(2,\mathbb{R}) \overrightarrow{\oplus}_R \mathfrak{g}_{6,85}$

- Describing representation: $R = D_{\frac{1}{2}} \oplus 4D_0$

- Structure tensor:

$$C_{12}^2 = 2, \quad C_{13}^3 = -2, \quad C_{23}^1 = 1, \quad C_{14}^4 = 1, \quad C_{15}^5 = -1, \quad C_{25}^4 = 1,$$
$$C_{34}^5 = 1, \quad C_{45}^6 = \varepsilon, \quad C_{49}^4 = 1, \quad C_{59}^5 = 1, \quad C_{69}^6 = 2, \quad C_{79}^7 = 1,$$
$$C_{89}^7 = 1, \quad C_{89}^8 = 1, \quad C_{78}^6 = 1.$$

- $\mathrm{codim}_{\mathfrak{g}}\,[\mathfrak{g},\mathfrak{g}] = 1.$

- $\dim Der(\mathfrak{g}) = 10.$

- $\dim O(\mathfrak{g}) = 71.$

- Rank $A(\mathfrak{g}) = 8$

- Conditions on invariants:

$$\frac{\partial F}{\partial x_j} = 0, \; j = 7,8,9.$$

- Codimension one subalgebra:

$$L\langle X_1, .., X_8\rangle \simeq L_{8,2}$$

- Invariants of subalgebra:

$$J_1 = 2\varepsilon x_2 x_3 x_8 + x_2 x_5^2 + x_1 x_4 x_5 - x_3 x_4^2 + \frac{\varepsilon}{2} x_1^2 x_8,$$
$$J_2 = x_6.$$

- Semi-invariance conditions:

$$\widehat{X}_9(J_1) = -2J_1, \; \widehat{X}_9(J_2) = -2J_2.$$

- Invariants of $L_{9,26}^{\varepsilon}$

$$I_1 = \frac{J_1}{J_2}.$$

$\mathbf{L}^p_{9,27}$

- Levi decomposition: $L^p_{9,27} = \mathfrak{sl}(2,\mathbb{R}) \overrightarrow{\oplus}_R \mathfrak{g}_{6,89}$

- Describing representation: $R = D_{\frac{1}{2}} \oplus 4D_0$

- Structure tensor:

$$
\begin{array}{llllll}
C^2_{12} = 2, & C^3_{13} = -2, & C^1_{23} = 1, & C^4_{14} = 1, & C^5_{15} = -1, & C^4_{25} = 1, \\
C^5_{34} = 1, & C^6_{45} = 1, & C^4_{49} = p, & C^5_{59} = p, & C^6_{69} = 2p, & C^7_{79} = p, \\
C^8_{79} = -1, & C^7_{89} = 1, & C^8_{89} = p, & C^6_{78} = 1.
\end{array}
$$

- $\mathrm{codim}_\mathfrak{g}[\mathfrak{g},\mathfrak{g}] = 1$.

- $\dim Der(\mathfrak{g}) = 10$.

- $\dim O(\mathfrak{g}) = 71$.

- Rank $A(\mathfrak{g}) = \begin{cases} 8 & \text{if } p \neq 0 \\ 6 & \text{if } p = 0 \end{cases}$

- Conditions on invariants:

$$
\frac{\partial F}{\partial x_j} = 0, \ j = 7,8,9 \text{ if } p \neq 0.
$$

- Codimension one subalgebra:

$$
L\langle X_1,..,X_8\rangle \simeq L_{8,2} \text{ if } p \neq 0.
$$

- Invariants of subalgebra:

$$
J_1 = 2x_2x_3x_8 + x_2x_5^2 + x_1x_4x_5 - x_3x_4^2 + \frac{1}{2}x_1^2x_8,
$$
$$
J_2 = x_6.
$$

- Semi-invariance conditions:

$$
\widehat{X}_9(J_1) = -2pJ_1, \ \widehat{X}_9(J_2) = -2pJ_2.
$$

- Invariants of $L_{9,27}^{p}$

$$I_1 = \frac{J_1}{J_2}, \; p \neq 0$$

$$I_1 = J_1, \; I_2 = J_2, \; I_3 = 2x_6x_9 + x_7^2 + x_8^2, \; p = 0.$$

$\mathbf{L}_{9,28}^{p,q} \; [pq \neq 0]$

- Levi decomposition: $L_{9,28}^{p,q} = \mathfrak{sl}\,(2,\mathbb{R}) \overrightarrow{\oplus}_R N_{6,1}^{p,q,p,q}$

- Describing representation: $R = D_{\frac{1}{2}} \oplus 4D_0$

- Structure tensor:

$$\begin{array}{llllll} C_{12}^2 = 2, & C_{13}^3 = -2, & C_{23}^1 = 1, & C_{14}^4 = 1, & C_{15}^5 = -1, & C_{25}^4 = 1, \\ C_{34}^5 = 1, & C_{48}^4 = p, & C_{49}^4 = q, & C_{58}^5 = p, & C_{59}^5 = q, & C_{68}^6 = 1, \\ C_{79}^7 = 1. \end{array}$$

- $\operatorname{codim}_{\mathfrak{g}}[\mathfrak{g},\mathfrak{g}] = 2.$

- $\dim Der\,(\mathfrak{g}) = 10.$

- $\dim O\,(\mathfrak{g}) = 71.$

- $\operatorname{Rank} A\,(\mathfrak{g}) = 8$

- Conditions on invariants:

$$\frac{\partial F}{\partial x_j} = 0, \; j = 8,9.$$

- Codimension two subalgebra:

$$L\,\langle X_1,..,X_7\rangle \simeq L_{5,1} \oplus 2L_1$$

- Invariants of subalgebra:

$$\begin{aligned} J_1 &= -x_2x_5^2 - x_1x_4x_5 + x_3x_4^2, \\ J_2 &= x_6, \\ J_3 &= x_7. \end{aligned}$$

- Semi-invariance conditions:

$$\widehat{X}_8\left(J_1\right) = -2pJ_1, \ \widehat{X}_8\left(J_2\right) = -J_2,$$
$$\widehat{X}_9\left(J_1\right) = -2qJ_1, \ \widehat{X}_9\left(J_3\right) = -J_3.$$

- Invariants of $L_{9,28}^{p,q}$

$$I_1 = \frac{J_1}{J_2^{2p}J_3^{2q}}.$$

$\mathbf{L}_{9,29}^{p}$

- Levi decomposition: $L_{9,29}^{p} = \mathfrak{sl}\left(2,\mathbb{R}\right) \overrightarrow{\oplus}_R N_{6,2}^{1,p,p}$

- Describing representation: $R = D_{\frac{1}{2}} \oplus 4D_0$

- Structure tensor:

$$\begin{array}{llllll}
C_{12}^2 = 2, & C_{13}^3 = -2, & C_{23}^1 = 1, & C_{14}^4 = 1, & C_{15}^5 = -1, & C_{25}^4 = 1, \\
C_{34}^5 = 1, & C_{48}^4 = 1, & C_{49}^4 = p, & C_{58}^5 = 1, & C_{59}^5 = p, & C_{69}^6 = 1, \\
C_{78}^6 = 1, & C_{79}^7 = 1. & & & &
\end{array}$$

- $\mathrm{codim}_{\mathfrak{g}}\left[\mathfrak{g},\mathfrak{g}\right] = 2.$

- $\dim Der\left(\mathfrak{g}\right) = 10.$

- $\dim O\left(\mathfrak{g}\right) = 71.$

- Rank $A\left(\mathfrak{g}\right) = 8$

- Conditions on invariants:

$$\frac{\partial F}{\partial x_j} = 0, \ j = 8,9.$$

- Codimension two subalgebra:

$$L\left\langle X_1,..,X_7\right\rangle \simeq L_{5,1} \oplus 2L_1$$

- Invariants of subalgebra:

$$J_1 = -x_2 x_5^2 - x_1 x_4 x_5 + x_3 x_4^2,$$
$$J_2 = x_6,$$
$$J_3 = x_7.$$

- Semi-invariance conditions:

$$\widehat{X}_8 (J_1) = -2J_1, \ \widehat{X}_8 (J_3) = -J_2,$$
$$\widehat{X}_9 (J_1) = -2qJ_1, \ \widehat{X}_9 (J_2) = -J_2, \ \widehat{X}_9 (J_3) = -J_3.$$

- Invariants of $L_{9,29}^p$

$$I_1 = \ln \left(\frac{J_1}{J_2^{2q}} \right) - 2J_3 J_2^{-1}.$$

$\mathbf{L}_{9,30}^{p,q} \ [p^2 + q^2 \neq 0]$

- Levi decomposition: $L_{9,30}^{p,q} = \mathfrak{sl}(2,\mathbb{R}) \overrightarrow{\oplus}_R N_{6,13}^{p,q,p,q}$

- Describing representation: $R = D_{\frac{1}{2}} \oplus 4D_0$

- Structure tensor:

$$\begin{aligned}
&C_{12}^2 = 2, \quad C_{13}^3 = -2, \quad C_{23}^1 = 1, \quad C_{14}^4 = 1, \quad C_{15}^5 = -1, \quad C_{25}^4 = 1, \\
&C_{34}^5 = 1, \quad C_{48}^4 = q, \quad C_{49}^4 = p, \quad C_{58}^5 = q, \quad C_{59}^5 = p, \quad C_{68}^6 = 1, \\
&C_{69}^7 = -1, \quad C_{78}^7 = 1, \quad C_{79}^6 = 1.
\end{aligned}$$

- $\text{codim}_{\mathfrak{g}} [\mathfrak{g}, \mathfrak{g}] = 2$.

- $\dim Der (\mathfrak{g}) = 10$.

- $\dim O (\mathfrak{g}) = 71$.

- Rank $A (\mathfrak{g}) = 8$

- Conditions on invariants:

$$\frac{\partial F}{\partial x_j} = 0, \ j = 8,9.$$

- Codimension two subalgebra:

$$L\langle X_1,...,X_7\rangle \simeq L_{5,1} \oplus 2L_1$$

- Invariants of subalgebra:

$$J_1 = -x_2 x_5^2 - x_1 x_4 x_5 + x_3 x_4^2,$$
$$J_2 = x_6,$$
$$J_3 = x_7.$$

- Semi-invariance conditions:

$$\widehat{X}_8(J_1) = -2qJ_1, \ \widehat{X}_8\left(J_3 J_2^{-1}\right) = 0,$$
$$\widehat{X}_9(J_1) = -2pJ_1, \ \widehat{X}_9\left(J_3 J_2^{-1}\right) = -\left(1 + \left(J_3 J_2^{-1}\right)^2\right).$$

- Invariants of $L_{9,30}^{p,q}$

$$I_1 = \ln\left(\frac{J_1}{(J_2^2 + J_3^2)^q}\right) - 2p\arctan\left(J_3 J_2^{-1}\right).$$

$\mathbf{L}_{9,31}$

- Levi decomposition: $L_{9,31} = \mathfrak{sl}(2,\mathbb{R}) \overrightarrow{\oplus}_R N_{6,20}^{1,0}$

- Describing representation: $R = D_{\frac{1}{2}} \oplus 4D_0$

- Structure tensor:

$$C_{12}^2 = 2, \ C_{13}^3 = -2, \ C_{23}^1 = 1, \ C_{14}^4 = 1, \ C_{15}^5 = -1, \ C_{25}^4 = 1,$$
$$C_{34}^5 = 1, \ C_{48}^4 = 1, \ C_{58}^5 = 1, \ C_{69}^6 = 1, \ C_{89}^7 = 1.$$

- $\text{codim}_{\mathfrak{g}}[\mathfrak{g},\mathfrak{g}] = 2.$

- $\dim Der(\mathfrak{g}) = 10.$

- $\dim O(\mathfrak{g}) = 71.$

- Rank $A(\mathfrak{g}) = 8$

- Conditions on invariants:

$$\frac{\partial F}{\partial x_j} = 0, \ j \neq 7.$$

- Invariants of $L_{9,31}$

$$I_1 = x_7.$$

$\mathbf{L}_{9,32}^{p;q} \ [pq \neq 0]$

- Levi decomposition: $L_{9,32}^{p;q} = \mathfrak{sl}(2,\mathbb{R}) \overrightarrow{\oplus}_R \mathfrak{g}_{6,1}$

- Describing representation: $R = D_1 \oplus 3D_0$

- Structure tensor:

$$
\begin{array}{llllll}
C_{12}^2 = 2, & C_{13}^3 = -2, & C_{23}^1 = 1, & C_{14}^4 = 2, & C_{16}^6 = -2, & C_{25}^4 = 2, \\
C_{26}^5 = 1, & C_{34}^5 = 1, & C_{35}^6 = 2, & C_{49}^4 = 1, & C_{59}^5 = 1, & C_{69}^6 = 1, \\
C_{79}^7 = p, & C_{89}^8 = q. & & & &
\end{array}
$$

- $\mathrm{codim}_{\mathfrak{g}}\,[\mathfrak{g},\mathfrak{g}] = 1.$

- $\dim Der(\mathfrak{g}) = \left\{ \begin{array}{ll} 11, & p \neq q \\ 13, & p = q \end{array} \right. .$

- $\dim O(\mathfrak{g}) = \left\{ \begin{array}{ll} 70, & p \neq q \\ 68, & p = q \end{array} \right. .$

- Rank $A(\mathfrak{g}) = 6$

- Conditions on invariants:

$$\frac{\partial F}{\partial x_9} = 0.$$

- Codimension one subalgebra:

$$L\langle X_1,..,X_8 \rangle \simeq L_{6,4} \oplus 2L_1$$

- Invariants of subalgebra:

$$J_1 = x_1 x_5 + 2x_2 x_6 - 2x_3 x_4,$$
$$J_2 = x_5^2 - 4x_4 x_6,$$
$$J_3 = x_7,$$
$$J_4 = x_8.$$

- Semi-invariance conditions:

$$\widehat{X}_9\left(J_1\right) = -J_1,\ \widehat{X}_8\left(J_2\right) = -2J_2,$$
$$\widehat{X}_9\left(J_3\right) = -pJ_3,\ \widehat{X}_9\left(J_4\right) = -qJ_4.$$

- Invariants of $L_{9,32}^{p,q}$

$$I_1 = \frac{J_1^2}{J_2},\ I_2 = \frac{J_2^p}{J_3^2},\ I_3 = \frac{J_3^q}{J_4^p}.$$

$\mathbf{L}_{9,33}^{p}$

- Levi decomposition: $L_{9,33}^{p} = \mathfrak{sl}\left(2,\mathbb{R}\right) \overrightarrow{\oplus}_R \mathfrak{g}_{6,2}$

- Describing representation: $R = D_1 \oplus 3D_0$

- Structure tensor:

$$C_{12}^2 = 2,\quad C_{13}^3 = -2,\quad C_{23}^1 = 1,\quad C_{14}^4 = 2,\quad C_{16}^6 = -2,\quad C_{25}^4 = 2,$$
$$C_{26}^5 = 1,\quad C_{34}^5 = 1,\quad C_{35}^6 = 2,\quad C_{49}^4 = 1,\quad C_{59}^5 = 1,\quad C_{69}^6 = 1,$$
$$C_{79}^7 = p,\quad C_{89}^8 = p,\quad C_{89}^7 = 1.$$

- $\operatorname{codim}_{\mathfrak{g}}\left[\mathfrak{g},\mathfrak{g}\right] = \left\{ \begin{array}{ll} 1, & p \neq 0 \\ 2, & p = 0 \end{array} \right.$.

- $\dim Der\left(\mathfrak{g}\right) = 11$.

- $\dim O\left(\mathfrak{g}\right) = 70$.

- Rank $A\left(\mathfrak{g}\right) = 6$

- Conditions on invariants:

$$\frac{\partial F}{\partial x_9} = 0.$$

- Codimension one subalgebra:

$$L\langle X_1, .., X_8\rangle \simeq L_{6,4} \oplus 2L_1$$

- Invariants of subalgebra:

$$J_1 = x_1 x_5 + 2x_2 x_6 - 2x_3 x_4,$$
$$J_2 = x_5^2 - 4x_4 x_6,$$
$$J_3 = x_7,$$
$$J_4 = x_8.$$

- Semi-invariance conditions:

$$\widehat{X}_9(J_1) = -J_1, \ \widehat{X}_8(J_2) = -2J_2,$$
$$\widehat{X}_9(J_3) = -pJ_3, \ \widehat{X}_9(J_4) = -J_3 - pJ_4.$$

- Invariants of $L_{9,33}^p$

$$I_1 = \frac{J_1^2}{J_2}, \ I_2 = \frac{J_2^p}{J_3^2}, \ I_3 = \frac{pJ_4 - J_3 \ln(J_4)}{pJ_3}, \ p \neq 0$$

$$I_1 = \frac{J_1^2}{J_2}, \ I_2 = J_3, \ I_3 = \ln(J_2) - 2\frac{J_4}{J_3}, \ p = 0.$$

$\mathbf{L}_{9,34}^{p;q} \ [p \neq 0, \ q \geq 0]$

- Levi decomposition: $L_{9,34}^{p;q} = \mathfrak{sl}(2,\mathbb{R}) \overrightarrow{\oplus}_R \mathfrak{g}_{6,8}$

- Describing representation: $R = D_1 \oplus 3D_0$

- Structure tensor:

$$C_{12}^2 = 2, \ \ C_{13}^3 = -2, \ \ C_{23}^1 = 1, \ \ C_{14}^4 = 2, \ \ C_{16}^6 = -2, \ \ C_{25}^4 = 2,$$
$$C_{26}^5 = 1, \ \ C_{34}^5 = 1, \ \ C_{35}^6 = 2, \ \ C_{49}^4 = p, \ \ C_{59}^5 = p, \ \ C_{69}^6 = p,$$
$$C_{79}^7 = q, \ \ C_{79}^8 = -1, \ \ C_{89}^7 = 1, \ \ C_{89}^8 = q.$$

- $\mathrm{codim}_{\mathfrak{g}} [\mathfrak{g}, \mathfrak{g}] = 1.$

- $\dim Der(\mathfrak{g}) = 11.$

- $\dim O(\mathfrak{g}) = 70.$

- Rank $A(\mathfrak{g}) = 6$

- Conditions on invariants:
$$\frac{\partial F}{\partial x_9} = 0.$$

- Codimension one subalgebra:
$$L\langle X_1,..,X_8 \rangle \simeq L_{6,4} \oplus 2L_1$$

- Invariants of subalgebra:
$$J_1 = x_1 x_5 + 2x_2 x_6 - 2x_3 x_4,$$
$$J_2 = x_5^2 - 4x_4 x_6,$$
$$J_3 = x_7,$$
$$J_4 = x_8.$$

- Semi-invariance conditions:
$$\widehat{X}_9(J_1) = -pJ_1, \ \widehat{X}_8(J_2) = -2pJ_2,$$
$$\widehat{X}_9(J_3) = -qJ_3 - J_4, \ \widehat{X}_9(J_4) = -J_3 - qJ_4.$$

- Invariants of $L_{9,34}^{p,q}$
$$I_1 = \frac{J_1^2}{J_2}, \ I_2 = \frac{J_2^q}{(J_3^2 + J_4^2)^p}, \ I_3 = (J_3^2 + J_4^2)\left(\frac{J_3 - iJ_4}{J_3 + iJ_4}\right)^{-iq}.$$

$\mathbf{L}_{9,35}^{p} \ [p \neq 0]$

- Levi decomposition: $L_{9,35}^{p} = \mathfrak{sl}(2,\mathbb{R}) \overrightarrow{\oplus}_R \mathfrak{g}_{6,1}$

- Describing representation: $R = D_{\frac{3}{2}} \oplus 2D_0$

- Structure tensor:

$$
\begin{array}{llllll}
C_{12}^2 = 2, & C_{13}^3 = -2, & C_{23}^1 = 1, & C_{14}^4 = 3, & C_{15}^5 = 1, & C_{16}^6 = -1, \\
C_{17}^7 = -3, & C_{25}^4 = 3, & C_{26}^5 = 2, & C_{27}^6 = 1, & C_{34}^5 = 1, & C_{35}^6 = 2, \\
C_{36}^7 = 3, & C_{49}^4 = 1, & C_{59}^5 = 1, & C_{69}^6 = 1, & C_{79}^7 = 1, & C_{89}^8 = p.
\end{array}
$$

- $\operatorname{codim}_{\mathfrak{g}} [\mathfrak{g}, \mathfrak{g}] = 1$.

- $\dim Der(\mathfrak{g}) = 10$.

- $\dim O(\mathfrak{g}) = 71$.

- Rank $A(\mathfrak{g}) = 8$

- Conditions on invariants:

$$
\frac{\partial F}{\partial x_j} = 0, \; j = 1, 2, 3, 9.
$$

- Codimension one subalgebra:

$$
L\langle X_1, .., X_8 \rangle \simeq L_{7,6} \oplus L_1
$$

- Invariants of subalgebra:

$$
J_1 = 27x_4^2 x_7^2 - 18x_4 x_5 x_6 x_7 - x_5^2 x_6^2 + 4\left(x_4 x_6^3 + x_5^3 x_7\right),
$$
$$
J_2 = x_8.
$$

- Semi-invariance conditions:

$$
\widehat{X}_9(J_1) = -4J_1, \; \widehat{X}_8(J_2) = -qJ_2.
$$

- Invariants of $L_{9,35}^p$

$$
I_1 = \frac{J_1^q}{J_2^4}.
$$

L$_{9,36}$

- Levi decomposition: $L_{9,36} = \mathfrak{sl}(2,\mathbb{R}) \overrightarrow{\oplus}_R \mathfrak{g}_{6,82}$

- Describing representation: $R = D_{\frac{3}{2}} \oplus 2D_0$

- Structure tensor:

$$C_{12}^2 = 2, \quad C_{13}^3 = -2, \quad C_{23}^1 = 1, \quad C_{14}^4 = 3, \quad C_{15}^5 = 1, \quad C_{16}^6 = -1,$$
$$C_{17}^7 = -3, \quad C_{25}^4 = 3, \quad C_{26}^5 = 2, \quad C_{27}^6 = 1, \quad C_{34}^5 = 1, \quad C_{35}^6 = 2,$$
$$C_{36}^7 = 3, \quad C_{49}^4 = 1, \quad C_{59}^5 = 1, \quad C_{69}^6 = 1, \quad C_{79}^7 = 1, \quad C_{89}^8 = 2,$$
$$C_{47}^8 = 1, \quad C_{56}^8 = -3.$$

- $\mathrm{codim}_{\mathfrak{g}}\,[\mathfrak{g},\mathfrak{g}] = 1.$

- $\dim Der\,(\mathfrak{g}) = 9.$

- $\dim O\,(\mathfrak{g}) = 71.$

- Rank $A\,(\mathfrak{g}) = 8$

- Conditions on invariants:

$$\frac{\partial F}{\partial x_9} = 0.$$

- Codimension one subalgebra:

$$L\langle X_1,..,X_8\rangle \simeq L_{8,19}$$

- Invariants of subalgebra:

$$J_1 = x_8, \quad J_2 = \sqrt{D},$$

where

$$D := \begin{vmatrix} 0 & 2x_2 & -2x_3 & 3x_4 & x_5 & -x_6 & -3x_7 & x_1 \\ -2x_2 & 0 & x_1 & 0 & 3x_4 & 2x_5 & x_6 & x_2 \\ 2x_3 & -x_1 & 0 & x_5 & 2x_6 & 3x_7 & 0 & x_3 \\ -3x_4 & 0 & -x_5 & 0 & 0 & 0 & x_8 & \frac{x_4}{2} \\ -x_5 & -3x_4 & -2x_6 & 0 & 0 & -3x_8 & 0 & \frac{x_5}{2} \\ x_6 & -3x_5 & -3x_7 & 0 & 3x_8 & 0 & 0 & \frac{x_6}{2} \\ 3x_7 & -x_6 & 0 & -x_8 & 0 & 0 & 0 & \frac{x_7}{2} \\ -x_1 & -x_2 & -x_3 & -\frac{x_4}{2} & -\frac{x_5}{2} & -\frac{x_6}{2} & -\frac{x_7}{2} & 0 \end{vmatrix}$$

72 R. Campoamor-Stursberg

- Semi-invariance conditions:

$$\widehat{X}_9\left(J_1\right) = -2J_1, \ \widehat{X}_8\left(J_2\right) = -J_2.$$

- Invariants of $L_{9,36}$

$$I_1 = \frac{J_2}{J_1^2}.$$

L$_{9,37}$

- Levi decomposition: $L_{9,37} = \mathfrak{sl}\left(2,\mathbb{R}\right) \overrightarrow{\oplus}_R \left(2\mathfrak{h}_1\right)$

- Describing representation: $R = 2D_{\frac{1}{2}} \oplus 2D_0$

- Structure tensor:

$$
\begin{array}{llllll}
C_{12}^2 = 2, & C_{13}^3 = -2, & C_{23}^1 = 1, & C_{14}^4 = 1, & C_{15}^5 = -1, & C_{16}^6 = 1, \\
C_{17}^7 = -1, & C_{25}^4 = 1, & C_{27}^6 = 1, & C_{34}^5 = 1, & C_{36}^7 = 1, & C_{45}^8 = 1, \\
C_{67}^9 = 1. & & & & &
\end{array}
$$

- $\operatorname{codim}_{\mathfrak{g}}\left[\mathfrak{g},\mathfrak{g}\right] = 0.$

- $\dim Der\left(\mathfrak{g}\right) = 9.$

- $\dim O\left(\mathfrak{g}\right) = 71.$

- Rank $A\left(\mathfrak{g}\right) = 6$

- Invariants of $L_{9,37}$:

$$
\begin{aligned}
I_1 &= x_8, \\
I_2 &= x_9, \\
I_3 &= \sqrt{D},
\end{aligned}
$$

where

$$D := \begin{vmatrix} 0 & 2x_2 & -2x_3 & x_4 & -x_5 & x_6 & -x_7 & x_1 \\ -2x_2 & 0 & x_1 & 0 & x_4 & 0 & x_6 & x_2 \\ 2x_3 & -x_1 & 0 & x_5 & 0 & x_7 & 0 & x_3 \\ -x_4 & 0 & -x_5 & 0 & x_8 & 0 & 0 & \frac{x_4}{2} \\ x_5 & -x_4 & 0 & -x_8 & 0 & 0 & 0 & \frac{x_5}{2} \\ -x_6 & 0 & -x_7 & 0 & 0 & 0 & x_9 & \frac{x_6}{2} \\ x_7 & -x_6 & 0 & 0 & 0 & -x_9 & 0 & \frac{x_7}{2} \\ -x_1 & -x_2 & -x_3 & -\frac{x_4}{2} & -\frac{x_5}{2} & -\frac{x_6}{2} & -\frac{x_7}{2} & 0 \end{vmatrix}$$

$L_{9,38}$

- Levi decomposition: $L_{9,38} = \mathfrak{sl}(2,\mathbb{R}) \overrightarrow{\oplus}_R (2A_{3,3})$

- Describing representation: $R = 2D_{\frac{1}{2}} \oplus 2D_0$

- Structure tensor:

$$\begin{aligned} C_{12}^2 &= 2, & C_{13}^3 &= -2, & C_{23}^1 &= 1, & C_{14}^4 &= 1, & C_{15}^5 &= -1, & C_{16}^6 &= 1, \\ C_{17}^7 &= -1, & C_{25}^4 &= 1, & C_{27}^6 &= 1, & C_{34}^5 &= 1, & C_{36}^7 &= 1, & C_{48}^4 &= 1, \\ C_{58}^5 &= 1, & C_{69}^6 &= 1, & C_{79}^7 &= 1. \end{aligned}$$

- $\mathrm{codim}_{\mathfrak{g}} [\mathfrak{g},\mathfrak{g}] = 2$.

- $\dim Der(\mathfrak{g}) = 9$.

- $\dim O(\mathfrak{g}) = 71$.

- Rank $A(\mathfrak{g}) = 8$

- Invariants of $L_{9,38}$

$$I_1 = \frac{x_4 x_7 (x_8 - x_9 + x_1) - 2x_3 x_4 x_6 + 2x_2 x_5 x_7 + x_5 x_6 (x_9 - x_8 + x_1)}{(x_4 x_7 - x_5 x_6)}.$$

$L_{9,39}$

- Levi decomposition: $L_{9,39} = \mathfrak{sl}(2,\mathbb{R}) \overrightarrow{\oplus}_R (\mathfrak{h}_1 \oplus \mathcal{A}_{3,3})$

- Describing representation: $R = 2D_{\frac{1}{2}} \oplus 2D_0$

- Structure tensor:

$$
\begin{array}{cccccc}
C_{12}^2 = 2, & C_{13}^3 = -2, & C_{23}^1 = 1, & C_{14}^4 = 1, & C_{15}^5 = -1, & C_{16}^6 = 1, \\
C_{17}^7 = -1, & C_{25}^4 = 1, & C_{27}^6 = 1, & C_{34}^5 = 1, & C_{36}^7 = 1, & C_{49}^4 = 1, \\
C_{59}^9 = 1, & C_{67}^8 = 1. & & & &
\end{array}
$$

- $\operatorname{codim}_{\mathfrak{g}}[\mathfrak{g}, \mathfrak{g}] = 1.$

- $\dim Der(\mathfrak{g}) = 10.$

- $\dim O(\mathfrak{g}) = 71.$

- $\operatorname{Rank} A(\mathfrak{g}) = 8$

- Conditions on invariants:

$$
\frac{\partial F}{\partial x_j} = 0, \ j \neq 8
$$

- Invariants of $L_{9,39}$

$$
I_1 = x_8.
$$

$L_{9,40}$

- Levi decomposition: $L_{9,40} = \mathfrak{sl}(2, \mathbb{R}) \overrightarrow{\oplus}_R (\mathcal{A}_{6,3})$

- Describing representation: $R = 2D_{\frac{1}{2}} \oplus 2D_0$

- Structure tensor:

$$
\begin{array}{cccccc}
C_{12}^2 = 2, & C_{13}^3 = -2, & C_{23}^1 = 1, & C_{14}^4 = 1, & C_{15}^5 = -1, & C_{16}^6 = 1, \\
C_{17}^7 = -1, & C_{25}^4 = 1, & C_{27}^6 = 1, & C_{34}^5 = 1, & C_{36}^7 = 1, & C_{67}^8 = 1, \\
C_{69}^4 = 1, & C_{79}^5 = 1. & & & &
\end{array}
$$

- $\operatorname{codim}_{\mathfrak{g}}[\mathfrak{g}, \mathfrak{g}] = 1.$

- $\dim Der(\mathfrak{g}) = 11.$

- $\dim O(\mathfrak{g}) = 70$.

- Rank $A(\mathfrak{g}) = 6$

- Invariants of $L_{9,40}$:

$$I_1 = x_8,$$
$$I_2 = x_4 x_7 - x_5 x_6 - x_8 x_9,$$
$$I_3 = \sqrt{D},$$

where

$$D := \begin{vmatrix} 0 & 2x_2 & -2x_3 & 3x_4 & x_5 & -x_6 & -3x_7 & x_1 \\ -2x_2 & 0 & x_1 & 0 & 3x_4 & 2x_5 & x_6 & x_2 \\ 2x_3 & -x_1 & 0 & x_5 & 2x_6 & 3x_7 & 0 & x_3 \\ -3x_4 & 0 & -x_5 & 0 & 0 & 0 & x_8 & \frac{x_4}{2} \\ -x_5 & -3x_4 & -2x_6 & 0 & 0 & -3x_8 & 0 & \frac{x_5}{2} \\ x_6 & -3x_5 & -3x_7 & 0 & 3x_8 & 0 & 0 & \frac{x_6}{2} \\ 3x_7 & -x_6 & 0 & -x_8 & 0 & 0 & 0 & \frac{x_7}{2} \\ -x_1 & -x_2 & -x_3 & -\frac{x_4}{2} & -\frac{x_5}{2} & -\frac{x_6}{2} & -\frac{x_7}{2} & 0 \end{vmatrix}$$

Remark 1 *Although here a fundamental system of invariants depends on all variables $\{x_1,..,x_9\}$, the functions I_1 and I_3 are actually a fundamental system of invariants of the subalgebra $L\langle X_1,..,X_8\rangle \simeq \mathfrak{sl}(2,\mathbb{R}) \overrightarrow{\oplus}_{2D_{\frac{1}{2}}\oplus D_0}(\mathfrak{h}_1 \oplus 2L_1)$.*

L$_{9,41}$

- Levi decomposition: $L_{9,41} = \mathfrak{sl}(2,\mathbb{R}) \overrightarrow{\oplus}_R \mathcal{A}_{6,4}$

- Describing representation: $R = 2D_{\frac{1}{2}} \oplus 2D_0$

- Structure tensor:

$$\begin{aligned}
&C_{12}^2 = 2, \quad C_{13}^3 = -2, \quad C_{23}^1 = 1, \quad C_{14}^4 = 1, \quad C_{15}^5 = -1, \quad C_{16}^6 = 1, \\
&C_{17}^7 = -1, \quad C_{25}^4 = 1, \quad\quad C_{27}^6 = 1, \quad C_{34}^5 = 1, \quad C_{36}^7 = 1, \quad\quad C_{47}^8 = 1, \\
&C_{56}^8 = -1, \quad C_{67}^9 = 1.
\end{aligned}$$

- $\mathrm{codim}_{\mathfrak{g}}[\mathfrak{g},\mathfrak{g}] = 0$.

- $\dim Der(\mathfrak{g}) = 10$.

- $\dim O(\mathfrak{g}) = 71$.

- Rank $A(\mathfrak{g}) = 6$

- Invariants of $L_{9,41}$:

$$I_1 = x_8,$$
$$I_2 = x_9,$$
$$I_3 = \sqrt{D},$$

where

$$D := \begin{vmatrix}
0 & 2x_2 & -2x_3 & x_4 & -x_5 & x_6 & -x_7 & x_1 \\
-2x_2 & 0 & x_1 & 0 & x_4 & 0 & x_6 & x_2 \\
2x_3 & -x_1 & 0 & x_5 & 0 & x_7 & 0 & x_3 \\
-x_4 & 0 & -x_5 & 0 & 0 & 0 & x_8 & \frac{x_4}{2} \\
x_5 & -x_4 & 0 & 0 & 0 & -x_8 & 0 & \frac{x_5}{2} \\
-x_6 & 0 & -x_7 & 0 & x_8 & 0 & x_9 & \frac{x_6}{2} \\
x_7 & -x_6 & 0 & -x_8 & 0 & -x_9 & 0 & \frac{x_7}{2} \\
-x_1 & -x_2 & -x_3 & -\frac{x_4}{2} & -\frac{x_5}{2} & -\frac{x_6}{2} & -\frac{x_7}{2} & 0
\end{vmatrix}.$$

$\mathbf{L_{9,42}}$

- Levi decomposition: $L_{9,42} = \mathfrak{sl}(2,\mathbb{R}) \overrightarrow{\oplus}_R \mathcal{A}_{6,5}$

- Describing representation: $R = 2D_{\frac{1}{2}} \oplus 2D_0$

- Structure tensor:

$$C_{12}^2 = 2, \quad C_{13}^3 = -2, \quad C_{23}^1 = 1, \quad C_{14}^4 = 1, \quad C_{15}^5 = -1, \quad C_{16}^6 = 1,$$
$$C_{17}^7 = -1, \quad C_{25}^4 = 1, \quad C_{27}^6 = 1, \quad C_{34}^5 = 1, \quad C_{36}^7 = 1, \quad C_{45}^9 = -1,$$
$$C_{47}^8 = 1, \quad C_{56}^8 = -1, \quad C_{67}^9 = 1.$$

- $\mathrm{codim}_{\mathfrak{g}}[\mathfrak{g},\mathfrak{g}] = 0$.

- $\dim Der(\mathfrak{g}) = 9$.

- $\dim O(\mathfrak{g}) = 72$.

- Rank $A(\mathfrak{g}) = 6$

- Invariants of $L_{9,42}$:

$$I_1 = x_8,$$
$$I_2 = x_9,$$
$$I_3 = \sqrt{D},$$

where

$$D := \begin{vmatrix} 0 & 2x_2 & -2x_3 & x_4 & -x_5 & x_6 & -x_7 & x_1 \\ -2x_2 & 0 & x_1 & 0 & x_4 & 0 & x_6 & x_2 \\ 2x_3 & -x_1 & 0 & x_5 & 0 & x_7 & 0 & x_3 \\ -x_4 & 0 & -x_5 & 0 & -x_9 & 0 & x_8 & \frac{x_4}{2} \\ x_5 & -x_4 & 0 & x_9 & 0 & -x_8 & 0 & \frac{x_5}{2} \\ -x_6 & 0 & -x_7 & 0 & x_8 & 0 & x_9 & \frac{x_6}{2} \\ x_7 & -x_6 & 0 & -x_8 & 0 & -x_9 & 0 & \frac{x_7}{2} \\ -x_1 & -x_2 & -x_3 & -\frac{x_4}{2} & -\frac{x_5}{2} & -\frac{x_6}{2} & -\frac{x_7}{2} & 0 \end{vmatrix}.$$

$\mathbf{L}_{9,43}^{p,q} \ [q \neq 0, \ |p| \leq 1]$

- Levi decomposition: $L_{9,43}^{p,q} = \mathfrak{sl}(2,\mathbb{R}) \overrightarrow{\oplus}_R \mathfrak{g}_{6,1}$

- Describing representation: $R = 2D_{\frac{1}{2}} \oplus 2D_0$

- Structure tensor:

$$\begin{array}{llllll} C_{12}^2 = 2, & C_{13}^3 = -2, & C_{23}^1 = 1, & C_{14}^4 = 1, & C_{15}^5 = -1, & C_{16}^6 = 1, \\ C_{17}^7 = -1, & C_{25}^4 = 1, & C_{27}^6 = 1, & C_{34}^5 = 1, & C_{36}^7 = 1, & C_{49}^4 = 1, \\ C_{59}^5 = 1, & C_{69}^6 = p, & C_{79}^7 = p, & C_{89}^8 = q. & & \end{array}$$

- $\operatorname{codim}_{\mathfrak{g}}[\mathfrak{g},\mathfrak{g}] = 1$.

- $\dim Der(\mathfrak{g}) = \begin{cases} 11, & p \neq 1 \\ 13, & p = 1 \end{cases}.$

- $\dim O(\mathfrak{g}) = \begin{cases} 70, & p \neq 1 \\ 68, & p = 1 \end{cases}$.

- Rank $A(\mathfrak{g}) = 8$

- Conditions on invariants:

$$\frac{\partial F}{\partial x_j} = 0, \; j = 1,2,3,9 \text{ if } p+1 \neq 0$$

$$\frac{\partial F}{\partial x_j} = 0, \; j = 1,2,3,8,9 \text{ if } p+1 = 0$$

- Codimension one subalgebra

$$L\langle X_1,..,X_8\rangle \simeq L_{7,7} \oplus L_1$$

- Invariants of subalgebra:

$$J_1 = x_4 x_7 - x_5 x_6$$
$$J_2 = x_8$$

- Semi-invariance conditions:

$$\widehat{X}_9(J_1) = -(1+p)J_1,$$
$$\widehat{X}_9(J_2) = -qJ_2.$$

- Invariants of $L_{9,43}^{p,q}$:

$$I_1 = \frac{J_1^q}{J_2^{1+p}}.$$

$\mathbf{L}_{9,44}^p$

- Levi decomposition: $L_{9,44}^p = \mathfrak{sl}(2,\mathbb{R}) \overrightarrow{\oplus}_R \mathfrak{g}_{6,6}$

- Describing representation: $R = 2D_{\frac{1}{2}} \oplus 2D_0$

- Structure tensor:

$$
\begin{array}{llllll}
C^2_{12} = 2, & C^3_{13} = -2, & C^1_{23} = 1, & C^4_{14} = 1, & C^5_{15} = -1, & C^6_{16} = 1, \\
C^7_{17} = -1, & C^4_{25} = 1, & C^6_{27} = 1, & C^5_{34} = 1, & C^7_{36} = 1, & C^4_{49} = p, \\
C^5_{59} = p, & C^4_{69} = 1, & C^6_{69} = p, & C^5_{79} = 1, & C^7_{79} = p, & C^8_{89} = 1.
\end{array}
$$

- $\operatorname{codim}_{\mathfrak{g}}[\mathfrak{g},\mathfrak{g}] = 1.$

- $\dim Der(\mathfrak{g}) = 11.$

- $\dim O(\mathfrak{g}) = 70.$

- Rank $A(\mathfrak{g}) = 8$

- Conditions on invariants:

$$
\frac{\partial F}{\partial x_j} = 0, \; j = 1, 2, 3, 9
$$

- Codimension one subalgebra

$$
L\langle X_1, .., X_8 \rangle \simeq L_{7,7} \oplus L_1
$$

- Invariants of subalgebra:

$$
J_1 = x_4 x_7 - x_5 x_6
$$
$$
J_2 = x_8
$$

- Semi-invariance conditions:

$$
\widehat{X}_9(J_1) = -2pJ_1,
$$
$$
\widehat{X}_9(J_2) = -J_2.
$$

- Invariants of $L^p_{9,44}$:

$$
I_1 = \frac{J_1}{J_2^{2p}}.
$$

$\mathbf{L}^{p,q}_{9,45} \; [q \neq 0]$

- Levi decomposition: $L_{9,45}^{p,q} = \mathfrak{sl}(2,\mathbb{R}) \overrightarrow{\oplus}_R \mathfrak{g}_{6,11}$

- Describing representation: $R = 2D_{\frac{1}{2}} \oplus 2D_0$

- Structure tensor:

$$
\begin{aligned}
&C_{12}^2 = 2, &&C_{13}^3 = -2, &&C_{23}^1 = 1, &&C_{14}^4 = 1, &&C_{15}^5 = -1, &&C_{16}^6 = 1, \\
&C_{17}^7 = -1, &&C_{25}^4 = 1, &&C_{27}^6 = 1, &&C_{34}^5 = 1, &&C_{36}^7 = 1, &&C_{49}^4 = p, \\
&C_{49}^6 = -1, &&C_{59}^5 = p, &&C_{59}^7 = -1, &&C_{69}^4 = 1, &&C_{69}^6 = p, &&C_{79}^5 = 1, \\
&C_{79}^7 = p, &&C_{89}^8 = q.
\end{aligned}
$$

- $\mathrm{codim}_\mathfrak{g}\,[\mathfrak{g},\mathfrak{g}] = 1.$

- $\dim Der\,(\mathfrak{g}) = 11.$

- $\dim O\,(\mathfrak{g}) = 70.$

- Rank $A\,(\mathfrak{g}) = 8$

- Conditions on invariants:

$$\frac{\partial F}{\partial x_j} = 0, \ j = 1,2,3,9$$

- Codimension one subalgebra

$$L\,\langle X_1,..,X_8 \rangle \simeq L_{7,7} \oplus L_1$$

- Invariants of subalgebra:

$$
\begin{aligned}
J_1 &= x_4 x_7 - x_5 x_6 \\
J_2 &= x_8
\end{aligned}
$$

- Semi-invariance conditions:

$$
\begin{aligned}
\widehat{X}_9\,(J_1) &= -2p J_1, \\
\widehat{X}_9\,(J_2) &= -q J_2.
\end{aligned}
$$

- Invariants of $L_{9,45}^{p,q}$:

$$I_1 = \frac{J_1^q}{J_2^{2p}}.$$

$\mathbf{L}_{9,46}^{p}$

- Levi decomposition: $L_{9,46}^{p} = \mathfrak{sl}(2,\mathbb{R}) \overrightarrow{\oplus}_R \mathfrak{g}_{6,13}$

- Describing representation: $R = 2D_{\frac{1}{2}} \oplus 2D_0$

- Structure tensor:

$$\begin{array}{llllll}
C_{12}^2 = 2, & C_{13}^3 = -2, & C_{23}^1 = 1, & C_{14}^4 = 1, & C_{15}^5 = -1, & C_{16}^6 = 1, \\
C_{17}^7 = -1, & C_{25}^4 = 1, & C_{27}^6 = 1, & C_{34}^5 = 1, & C_{36}^7 = 1, & C_{49}^4 = 1, \\
C_{45}^8 = 1, & C_{59}^5 = 1, & C_{69}^6 = p, & C_{79}^7 = p, & C_{89}^8 = 2.
\end{array}$$

- $\operatorname{codim}_{\mathfrak{g}}[\mathfrak{g},\mathfrak{g}] = 1.$

- $\dim Der(\mathfrak{g}) = \begin{cases} 10, & p \neq 1 \\ 11, & p = 1 \end{cases}.$

- $\dim O(\mathfrak{g}) = \begin{cases} 71, & p \neq 1 \\ 70, & p = 1 \end{cases}.$

- Rank $A(\mathfrak{g}) = 8$

- Conditions on invariants:

$$\frac{\partial F}{\partial x_9} = 0.$$

- Codimension one subalgebra

$$L\langle X_1,..,X_8\rangle \simeq L_{8,13}^{\varepsilon=-1}$$

- Invariants of subalgebra:

$$J_1 = x_8 \quad J_2 = \sqrt{D},$$

where

$$
D := \begin{vmatrix}
0 & 2x_2 & -2x_3 & x_4 & -x_5 & x_6 & -x_7 & x_1 \\
-2x_2 & 0 & x_1 & 0 & x_4 & 0 & x_6 & x_2 \\
2x_3 & -x_1 & 0 & x_5 & 0 & x_7 & 0 & x_3 \\
-x_4 & 0 & -x_5 & 0 & x_8 & 0 & 0 & \frac{x_4}{2} \\
x_5 & -x_4 & 0 & -x_8 & 0 & 0 & 0 & \frac{x_5}{2} \\
-x_6 & 0 & -x_7 & 0 & 0 & 0 & 0 & \frac{x_6}{2} \\
x_7 & -x_6 & 0 & 0 & 0 & 0 & 0 & \frac{x_7}{2} \\
-x_1 & -x_2 & -x_3 & -\frac{x_4}{2} & -\frac{x_5}{2} & -\frac{x_6}{2} & -\frac{x_7}{2} & 0
\end{vmatrix} .
$$

- Semi-invariance conditions:

$$
\widehat{X}_9\left(J_1\right) = -2J_1,
$$
$$
\widehat{X}_9\left(J_2\right) = -2\left(p+1\right)J_2.
$$

- Invariants of $L^p_{9,46}$:

$$
I_1 = \frac{J_2}{J_2^{1+p}}.
$$

$L_{9,47}$

- Levi decomposition: $L_{9,47} = \mathfrak{sl}\left(2,\mathbb{R}\right) \overrightarrow{\oplus}_R \mathfrak{g}_{6,15}$

- Describing representation: $R = 2D_{\frac{1}{2}} \oplus 2D_0$

- Structure tensor:

$$
\begin{aligned}
&C^2_{12} = 2, \quad C^3_{13} = -2, \quad C^1_{23} = 1, \quad C^4_{14} = 1, \quad C^5_{15} = -1, \quad C^6_{16} = 1, \\
&C^7_{17} = -1, \quad C^4_{25} = 1, \quad C^6_{27} = 1, \quad C^5_{34} = 1, \quad C^7_{36} = 1, \quad C^4_{49} = 1, \\
&C^8_{67} = 1, \quad C^5_{59} = 1, \quad C^4_{69} = 1, \quad C^6_{69} = 1, \quad C^5_{79} = 1, \quad C^7_{79} = 1, \\
&C^8_{89} = 2.
\end{aligned}
$$

- $\operatorname{codim}_{\mathfrak{g}}[\mathfrak{g},\mathfrak{g}] = 1$.

- $\dim Der\left(\mathfrak{g}\right) = 10$.

- $\dim O\left(\mathfrak{g}\right) = 71$.

- Rank $A(\mathfrak{g}) = 8$

- Conditions on invariants:

$$\frac{\partial F}{\partial x_9} = 0.$$

- Codimension one subalgebra

$$L\langle X_1,..,X_8\rangle \simeq L_{8,13}^{\varepsilon=-1}$$

- Invariants of subalgebra:

$$J_1 = x_8$$
$$J_2 = \sqrt{D},$$

where

$$D := \begin{vmatrix} 0 & 2x_2 & -2x_3 & x_4 & -x_5 & x_6 & -x_7 & x_1 \\ -2x_2 & 0 & x_1 & 0 & x_4 & 0 & x_6 & x_2 \\ 2x_3 & -x_1 & 0 & x_5 & 0 & x_7 & 0 & x_3 \\ -x_4 & 0 & -x_5 & 0 & 0 & 0 & 0 & \frac{x_4}{2} \\ x_5 & -x_4 & 0 & 0 & 0 & 0 & 0 & \frac{x_5}{2} \\ -x_6 & 0 & -x_7 & 0 & 0 & 0 & x_8 & \frac{x_6}{2} \\ x_7 & -x_6 & 0 & 0 & 0 & -x_8 & 0 & \frac{x_7}{2} \\ -x_1 & -x_2 & -x_3 & -\frac{x_4}{2} & -\frac{x_5}{2} & -\frac{x_6}{2} & -\frac{x_7}{2} & 0 \end{vmatrix}.$$

- Semi-invariance conditions:

$$\widehat{X}_9(J_1) = -2J_1,$$
$$\widehat{X}_9(J_2) = -4J_2.$$

- Invariants of $L_{9,47}$:

$$I_1 = \frac{J_2}{J_2^2}.$$

$L_{9,48}$

- Levi decomposition: $L_{9,48} = \mathfrak{sl}(2,\mathbb{R}) \overrightarrow{\oplus}_R \mathfrak{g}_{6,53}$

- Describing representation: $R = 2D_{\frac{1}{2}} \oplus 2D_0$

- Structure tensor:

$$
\begin{array}{llllll}
C_{12}^2 = 2, & C_{13}^3 = -2, & C_{23}^1 = 1, & C_{14}^4 = 1, & C_{15}^5 = -1, & C_{16}^6 = 1, \\
C_{17}^7 = -1, & C_{25}^4 = 1, & C_{27}^6 = 1, & C_{34}^5 = 1, & C_{36}^7 = 1, & C_{68}^4 = 1, \\
C_{69}^6 = 1, & C_{78}^5 = 1, & C_{79}^7 = 1, & C_{89}^8 = -1. &
\end{array}
$$

- $\mathrm{codim}_{\mathfrak{g}}\,[\mathfrak{g},\mathfrak{g}] = 1.$

- $\dim Der\,(\mathfrak{g}) = 10.$

- $\dim O\,(\mathfrak{g}) = 71.$

- Rank $A\,(\mathfrak{g}) = 8$

- Conditions on invariants:

$$\frac{\partial F}{\partial x_9} = 0.$$

- Codimension one subalgebra

$$L\,\langle X_1,..,X_8 \rangle \simeq L_{8,14}$$

- Invariants of subalgebra:

$$J_1 = x_1 x_4 x_5 + x_5 x_6 x_8 - x_4 x_7 x_8 + x_2 x_5^2 - x_3 x_4^2$$
$$J_2 = x_4 x_7 - x_5 x_6.$$

- Semi-invariance conditions:

$$\widehat{X}_9\,(J_1) = 0,$$
$$\widehat{X}_9\,(J_2) = -J_2.$$

- Invariants of $L_{9,48}$:

$$I_1 = J_1.$$

L$^p_{9,49}$

- Levi decomposition: $L^p_{9,49} = \mathfrak{sl}(2,\mathbb{R}) \overrightarrow{\oplus}_R \mathfrak{g}_{6,53}$

- Describing representation: $R = 2D_{\frac{1}{2}} \oplus 2D_0$

- Structure tensor:
$$
\begin{array}{llllll}
C^2_{12} = 2, & C^3_{13} = -2, & C^1_{23} = 1, & C^4_{14} = 1, & C^5_{15} = -1, & C^6_{16} = 1, \\
C^7_{17} = -1, & C^4_{25} = 1, & C^6_{27} = 1, & C^5_{34} = 1, & C^7_{36} = 1, & C^4_{49} = 1, \\
C^5_{59} = 1, & C^4_{68} = 1, & C^6_{69} = 1-p, & C^5_{78} = 1, & C^7_{79} = 1-p, & C^8_{89} = p.
\end{array}
$$

- $\mathrm{codim}_{\mathfrak{g}}\,[\mathfrak{g},\mathfrak{g}] = \left\{ \begin{array}{ll} 1, & p \neq 0 \\ 2, & p = 0 \end{array} \right.$.

- $\dim Der\,(\mathfrak{g}) = 10$.

- $\dim O\,(\mathfrak{g}) = 71$.

- Rank $A\,(\mathfrak{g}) = 8$

- Conditions on invariants:
$$
\frac{\partial F}{\partial x_9} = 0.
$$

- Codimension one subalgebra
$$
L\,\langle X_1,..,X_8 \rangle \simeq L_{8,14}
$$

- Invariants of subalgebra:
$$
J_1 = x_1 x_4 x_5 + x_5 x_6 x_8 - x_4 x_7 x_8 + x_2 x_5^2 - x_3 x_4^2
$$
$$
J_2 = x_4 x_7 - x_5 x_6.
$$

- Semi-invariance conditions:
$$
\widehat{X}_9\,(J_1) = -2J_1,
$$
$$
\widehat{X}_9\,(J_2) = (p-2)\,J_2.
$$

- Invariants of $L^p_{9,49}$:
$$
I_1 = J_1^{p-2} J_2^2.
$$

L$_{9,50}$

- Levi decomposition: $L_{9,50} = \mathfrak{sl}(2,\mathbb{R}) \overrightarrow{\oplus}_R \mathfrak{g}_{6,76}$

- Describing representation: $R = 2D_{\frac{1}{2}} \oplus 2D_0$

- Structure tensor:

$$
\begin{array}{llllll}
C_{12}^2 = 2, & C_{13}^3 = -2, & C_{23}^1 = 1, & C_{14}^4 = 1, & C_{15}^5 = -1, & C_{16}^6 = 1, \\
C_{17}^7 = -1, & C_{25}^4 = 1, & C_{27}^6 = 1, & C_{34}^5 = 1, & C_{36}^7 = 1, & C_{49}^4 = 3, \\
C_{59}^5 = 3, & C_{67}^8 = 1, & C_{68}^4 = 1, & C_{69}^6 = 1, & C_{78}^5 = 1, & C_{79}^7 = 1, \\
C_{89}^8 = 2.
\end{array}
$$

- $\mathrm{codim}_{\mathfrak{g}}[\mathfrak{g},\mathfrak{g}] = 1$.

- $\dim Der(\mathfrak{g}) = 9$.

- $\dim O(\mathfrak{g}) = 72$.

- Rank $A(\mathfrak{g}) = 8$

- Conditions on invariants:

$$\frac{\partial F}{\partial x_9} = 0.$$

- Codimension one subalgebra

$$L\langle X_1,..,X_8\rangle \simeq L_{8,15}$$

- Invariants of subalgebra:

$$J_1 = x_1 x_4 x_5 + x_5 x_6 x_8 - x_4 x_7 x_8 + x_2 x_5^2 - x_3 x_4^2 + \frac{1}{3} x_8^3$$

$$J_2 = x_4 x_7 - x_5 x_6 - \frac{1}{2} x_8^2.$$

- Semi-invariance conditions:

$$\widehat{X}_9(J_1) = -6J_1,$$
$$\widehat{X}_9(J_2) = -4J_2.$$

- Invariants of $L_{9,50}$:

$$I_1 = \frac{J_1^2}{J_2^3}.$$

$\mathbf{L}^p_{9,51}\ [\,|p| \leq 1\,]$

- Levi decomposition: $L^p_{9,51} = \mathfrak{sl}(2,\mathbb{R}) \overrightarrow{\oplus}_R \mathfrak{g}_{6,82}$

- Describing representation: $R = 2D_{\frac{1}{2}} \oplus 2D_0$

- Structure tensor:

$$\begin{array}{llllll}
C^2_{12} = 2, & C^3_{13} = -2, & C^1_{23} = 1, & C^4_{14} = 1, & C^5_{15} = -1, & C^6_{16} = 1, \\
C^7_{17} = -1, & C^4_{25} = 1, & C^6_{27} = 1, & C^5_{34} = 1, & C^7_{36} = 1, & C^4_{49} = 1, \\
C^8_{47} = 1, & C^8_{56} = -1, & C^5_{59} = 1, & C^6_{69} = p, & C^7_{79} = p, & C^8_{89} = 1+p.
\end{array}$$

- $\mathrm{codim}_{\mathfrak{g}}[\mathfrak{g},\mathfrak{g}] = 1.$

- $\dim Der(\mathfrak{g}) = 10.$

- $\dim O(\mathfrak{g}) = 71.$

- Rank $A(\mathfrak{g}) = \begin{cases} 8 & \text{if } p+1 \neq 0 \\ 6 & \text{if } p+1 = 0 \end{cases}$

- Conditions on invariants:

$$\frac{\partial F}{\partial x_9} = 0 \text{ if } p+1 \neq 0.$$

- Codimension one (and three if $p+1 = 0$) subalgebras:

$$L\langle X_1,..,X_8\rangle \simeq L_{8,13}$$
$$L\langle X_4,..,X_9\rangle \simeq \mathfrak{g}_{6,82} \text{ if } p+1 = 0$$

- Invariants of subalgebra:

$$J_1 = \left(x_1^2 + 4x_2x_3\right)x_8^2 + 2x_1x_8\left(x_4x_7 + x_5x_6\right) - 2x_4x_5x_6x_7 +$$
$$+ 4x_8\left(x_2x_5x_7 - x_3x_4x_6\right) + x_4^2x_7^2 + x_5^2x_6^2,$$
$$J_2 = x_8,$$
$$J_3 = x_4x_7 - x_5x_6 - x_8x_9 \text{ if } p+1 = 0.$$

- Semi-invariance conditions:

$$\widehat{X}_9\left(J_1\right) = -2\left(1+p\right)J_1,$$
$$\widehat{X}_9\left(J_2\right) = -\left(1+p\right)J_2,$$
$$\widehat{X}_9\left(J_3\right) = 0,\ \left(p+1=0\right).$$

- Invariants of $L_{9,51}^p$:

$$I_1 = \frac{J_1}{J_2^2}\ \text{if}\ p+1 \neq 0,$$
$$I_1 = J_1,\ I_2 = J_2,$$
$$I_3 = x_4x_7 - x_5x_6 - x_8x_9\ \text{if}\ p+1 = 0$$

L$_{9,52}$

- Levi decomposition: $L_{9,52} = \mathfrak{sl}\left(2,\mathbb{R}\right)\overrightarrow{\oplus}_R \mathfrak{g}_{6,82}$

- Describing representation: $R = 2D_{\frac{1}{2}} \oplus 2D_0$

- Structure tensor:

$$
\begin{array}{llllll}
C_{12}^2 = 2, & C_{13}^3 = -2, & C_{23}^1 = 1, & C_{14}^4 = 1, & C_{15}^5 = -1, & C_{16}^6 = 1, \\
C_{17}^7 = -1, & C_{25}^4 = 1, & C_{27}^6 = 1, & C_{34}^5 = 1, & C_{36}^7 = 1, & C_{45}^8 = 1, \\
C_{49}^4 = 1, & C_{59}^5 = 1, & C_{67}^8 = 1, & C_{69}^6 = 1, & C_{79}^7 = 1, & C_{89}^8 = 2.
\end{array}
$$

- $\operatorname{codim}_{\mathfrak{g}}\left[\mathfrak{g},\mathfrak{g}\right] = 1$.

- $\dim Der\left(\mathfrak{g}\right) = 10$.

- $\dim O\left(\mathfrak{g}\right) = 71$.

- $\operatorname{Rank} A\left(\mathfrak{g}\right) = 8$

- Conditions on invariants:

$$\frac{\partial F}{\partial x_9} = 0.$$

- Codimension one subalgebra

$$L\langle X_1,..,X_8\rangle \simeq L_{8,13}$$

- Invariants of subalgebra:

$$J_1 = \left(x_1^2 + 4x_2x_3\right)x_8^2 + 2x_1x_8\left(x_4x_7 + x_5x_6\right) - 2x_4x_5x_6x_7 +$$
$$+ 4x_8\left(x_2x_5x_7 - x_3x_4x_6\right) + x_4^2x_7^2 + x_5^2x_6^2,$$
$$J_2 = x_8.$$

- Semi-invariance conditions:

$$\widehat{X}_9\left(J_1\right) = -4J_1,$$
$$\widehat{X}_9\left(J_2\right) = -2J_2.$$

- Invariants of $L_{9,52}$:

$$I_1 = \frac{J_1}{J_2^2}.$$

$\mathbf{L}_{9,53}^p\ [p \geq 0]$

- Levi decomposition: $L_{9,53}^p = \mathfrak{sl}\left(2,\mathbb{R}\right)\overrightarrow{\oplus}_R\mathfrak{g}_{6,92}$

- Describing representation: $R = 2D_{\frac{1}{2}} \oplus 2D_0$

- Structure tensor:

$$
\begin{array}{llllll}
C_{12}^2 = 2, & C_{13}^3 = -2, & C_{23}^1 = 1, & C_{14}^4 = 1, & C_{15}^5 = -1, & C_{16}^6 = 1, \\
C_{17}^7 = -1, & C_{25}^4 = 1, & C_{27}^6 = 1, & C_{34}^5 = 1, & C_{36}^7 = 1, & C_{45}^8 = 1, \\
C_{49}^4 = p, & C_{49}^6 = -1, & C_{59}^5 = p, & C_{59}^7 = -1, & C_{67}^8 = 1, & C_{69}^4 = 1, \\
C_{69}^6 = p, & C_{79}^5 = 1, & C_{79}^7 = p, & C_{89}^8 = 2p.
\end{array}
$$

- $\mathrm{codim}_{\mathfrak{g}}\left[\mathfrak{g},\mathfrak{g}\right] = 1.$

- $\dim Der\left(\mathfrak{g}\right) = 10.$

- $\dim O\left(\mathfrak{g}\right) = 71.$

- Rank $A(\mathfrak{g}) = \begin{cases} 8 & \text{if } p \neq 0 \\ 6 & \text{if } p = 0 \end{cases}$

- Conditions on invariants:

$$\frac{\partial F}{\partial x_9} = 0.$$

- Codimension one subalgebra

$$L\langle X_1,..,X_8 \rangle \simeq L_{8,13}$$
$$L\langle X_4,..,X_9 \rangle \simeq \mathfrak{g}_{6,92} \text{ if } p = 0.$$

- Invariants of subalgebra:

$$J_1 = \left(x_1^2 + 4x_2x_3\right)x_8^2 + 2x_1x_8\left(x_4x_7 + x_5x_6\right) - 2x_4x_5x_6x_7 +$$
$$+ 4x_8\left(x_2x_5x_7 - x_3x_4x_6\right) + x_4^2x_7^2 + x_5^2x_6^2,$$
$$J_2 = x_8,$$
$$J_3 = x_4x_7 - x_5x_6 - x_8x_9 \text{ if } p = 0.$$

- Semi-invariance conditions:

$$\widehat{X}_9\left(J_1\right) = -4pJ_1,$$
$$\widehat{X}_9\left(J_2\right) = -2pJ_2,$$
$$\widehat{X}_9\left(J_3\right) = 0 \text{ if } p = 0.$$

- Invariants of $L_{9,53}^p$:

$$I_1 = \frac{J_1}{J_2^2} \text{ if } p \neq 0,$$
$$I_1 = J_1, I_2 = J_2, I_3 = J_3 \text{ if } p = 0.$$

$L_{9,54}$

- Levi decomposition: $L_{9,54} = \mathfrak{sl}(2,\mathbb{R}) \overrightarrow{\oplus}_R \mathcal{N}_{6,18}^{0,1,1}$

- Describing representation: $R = 2D_{\frac{1}{2}} \oplus 2D_0$

- Structure tensor:

$$
\begin{aligned}
&C^2_{12} = 2, \quad C^3_{13} = -2, \quad C^1_{23} = 1, \quad C^4_{14} = 1, \quad C^5_{15} = -1, \quad C^6_{16} = 1, \\
&C^7_{17} = -1, \quad C^4_{25} = 1, \quad C^6_{27} = 1, \quad C^5_{34} = 1, \quad C^7_{36} = 1, \quad C^4_{48} = 1, \\
&C^6_{49} = 1, \quad C^5_{58} = 1, \quad C^7_{59} = 1, \quad C^6_{68} = 1, \quad C^4_{69} = -1, \quad C^7_{78} = 1, \\
&C^5_{79} = -1.
\end{aligned}
$$

- $\operatorname{codim}_{\mathfrak{g}}[\mathfrak{g}, \mathfrak{g}] = 2$.

- $\dim Der(\mathfrak{g}) = 9$.

- $\dim O(\mathfrak{g}) = 72$.

- Rank $A(\mathfrak{g}) = 8$

- Conditions on invariants:

$$
\frac{\partial F}{\partial x_8} = 0.
$$

- Codimension one subalgebra

$$
L\langle X_1, .. X_7, X_9 \rangle \simeq L^0_{8,18}
$$

- Invariants of subalgebra:

$$
\begin{aligned}
J_1 &= -2x_1\left(x_6 x_7 + x_4 x_5\right) - 2x_9\left(x_4 x_7 - x_5 x_6\right) + 2x_3\left(x_4^2 + x_6^2\right) + \\
&\quad - 2x_2\left(x_5^2 + x_7^2\right), \\
J_2 &= x_4 x_7 - x_5 x_6.
\end{aligned}
$$

- Semi-invariance conditions:

$$
\begin{aligned}
\widehat{X}_9(J_1) &= -2J_1, \\
\widehat{X}_9(J_2) &= -2J_2.
\end{aligned}
$$

- Invariants of $L_{9,54}$:

$$
I_1 = \frac{J_1}{J_2}.
$$

$L_{9,55}$

- Levi decomposition: $L_{9,55} = \mathfrak{sl}(2,\mathbb{R}) \overrightarrow{\oplus}_R \mathfrak{g}_{6,1}$

- Describing representation: $R = D_2 \oplus D_0$

- Structure tensor:

$$\begin{array}{llllll}
C^2_{12} = 2, & C^3_{13} = -2, & C^1_{23} = 1, & C^4_{14} = 4, & C^5_{15} = 2, & C^7_{17} = -2, \\
C^8_{18} = -4, & C^4_{25} = 4, & C^5_{26} = 3, & C^6_{27} = 2, & C^7_{28} = 1, & C^5_{34} = 1, \\
C^6_{35} = 2, & C^7_{36} = 3, & C^8_{37} = 4, & C^4_{49} = 1, & C^5_{59} = 1, & C^6_{69} = 1, \\
C^7_{79} = 1, & C^8_{89} = 1.
\end{array}$$

- $\mathrm{codim}_{\mathfrak{g}}\,[\mathfrak{g},\mathfrak{g}] = 1.$

- $\dim Der(\mathfrak{g}) = 9.$

- $\dim O(\mathfrak{g}) = 72.$

- Rank $A(\mathfrak{g}) = 8$

- Conditions on invariants:

$$\frac{\partial F}{\partial x_j} = 0,\ j = 1,2,3,9.$$

- Codimension one subalgebra

$$L\langle X_1,..X_8 \rangle \simeq L_{8,21}$$

- Invariants of subalgebra:

$$J_1 = 3x_5x_7 - x_6^2 - 12x_4x_8,$$
$$J_2 = -2x_6^3 + 9x_5x_6x_7 + 72x_4x_6x_8 - 27\left(x_4x_7^2 + x_5^2x_8\right).$$

- Semi-invariance conditions:

$$\widehat{X}_9(J_1) = -2J_1,$$
$$\widehat{X}_9(J_2) = -3J_2.$$

- Invariants of $L_{9,55}$:

$$I_1 = \frac{J_1^3}{J_2^2}.$$

$\mathbf{L}_{9,56}^{p}$

- Levi decomposition: $L_{9,56}^{p} = \begin{cases} \mathfrak{sl}(2,\mathbb{R}) \overrightarrow{\oplus}_R \mathfrak{g}_{6,1}, & p \neq 0 \\ \mathfrak{sl}(2,\mathbb{R}) \overrightarrow{\oplus}_R \left(A_{4,5}^{1,1} \oplus 2L_1\right), & p = 0. \end{cases}$

- Describing representation: $R = D_1 \oplus D_{\frac{1}{2}} \oplus D_0$

- Structure tensor:

$$\begin{array}{llllll}
C_{12}^2 = 2, & C_{13}^3 = -2, & C_{23}^1 = 1, & C_{14}^4 = 2, & C_{16}^6 = -2, & C_{17}^7 = 1, \\
C_{18}^8 = -1, & C_{25}^4 = 2, & C_{26}^5 = 1, & C_{28}^7 = 1, & C_{34}^5 = 1, & C_{35}^6 = 2, \\
C_{37}^8 = 1, & C_{49}^4 = 1, & C_{59}^5 = 1, & C_{69}^6 = 1, & C_{79}^7 = p, & C_{89}^8 = p.
\end{array}$$

- $\mathrm{codim}_{\mathfrak{g}}\,[\mathfrak{g},\mathfrak{g}] = 1.$

- $\dim Der(\mathfrak{g}) = 10.$

- $\dim O(\mathfrak{g}) = 71.$

- Rank $A(\mathfrak{g}) = 8$

- Conditions on invariants:

$$\frac{\partial F}{\partial x_j} = 0,\ j = 1,2,3,9.$$

- Codimension one subalgebra

$$L\langle X_1,..X_8\rangle \simeq L_{8,22}$$

- Invariants of subalgebra:

$$J_1 = -4x_4x_6 + x_5^2,$$
$$J_2 = x_5x_7x_8 - x_4x_8^2 - x_6x_7^2.$$

- Semi-invariance conditions:

$$\widehat{X}_9\left(J_1\right) = -2J_1,$$
$$\widehat{X}_9\left(J_2\right) = -\left(1+2p\right)J_2.$$

- Invariants of $L_{9,56}^p$:

$$I_1 = \frac{J_1^{2p+1}}{J_2^2}.$$

$L_{9,57}$

- Levi decomposition: $L_{9,57} = \mathfrak{sl}\left(2,\mathbb{R}\right) \overrightarrow{\oplus}_R \left(3L_1 \oplus A_{3,3}\right)$

- Describing representation: $R = D_1 \oplus D_{\frac{1}{2}} \oplus D_0$

- Structure tensor:

$$\begin{array}{llllll}
C_{12}^2 = 2, & C_{13}^3 = -2, & C_{23}^1 = 1, & C_{14}^4 = 2, & C_{16}^6 = -2, & C_{17}^7 = 1, \\
C_{18}^8 = -1, & C_{25}^4 = 2, & C_{26}^5 = 1, & C_{28}^7 = 1, & C_{34}^5 = 1, & C_{35}^6 = 2, \\
C_{37}^8 = 1, & C_{79}^7 = 1, & C_{89}^8 = 1. & & &
\end{array}$$

- $\operatorname{codim}_{\mathfrak{g}}\left[\mathfrak{g},\mathfrak{g}\right] = 1$.

- $\dim Der\left(\mathfrak{g}\right) = 10$.

- $\dim O\left(\mathfrak{g}\right) = 71$.

- Rank $A\left(\mathfrak{g}\right) = 8$

- Conditions on invariants:

$$\frac{\partial F}{\partial x_j} = 0, \; j = 1,2,3,7,8,9.$$

- Codimension one subalgebra

$$L\left\langle X_1,..,X_8\right\rangle \simeq L_{8,22}$$

- Invariants of subalgebra:

$$J_1 = -4x_4x_6 + x_5^2,$$
$$J_2 = x_5x_7x_8 - x_4x_8^2 - x_6x_7^2.$$

- Semi-invariance conditions:

$$\widehat{X}_9(J_1) = 0,$$
$$\widehat{X}_9(J_2) = -2J_2.$$

- Invariants of $L_{9,57}$:

$$I_1 = J_1.$$

$\mathbf{L}_{9,58}$

- Levi decomposition: $L_{9,58} = \mathfrak{sl}(2,\mathbb{R}) \overrightarrow{\oplus}_R (\mathfrak{h}_1 \oplus 3L_1)$

- Describing representation: $R = D_1 \oplus D_{\frac{1}{2}} \oplus D_0$

- Structure tensor:

$$
\begin{array}{llllll}
C_{12}^2 = 2, & C_{13}^3 = -2, & C_{23}^1 = 1, & C_{14}^4 = 2, & C_{16}^6 = -2, & C_{17}^7 = 1, \\
C_{18}^8 = -1, & C_{25}^4 = 2, & C_{26}^5 = 1, & C_{28}^7 = 1, & C_{34}^5 = 1, & C_{35}^6 = 2, \\
C_{37}^8 = 1, & C_{78}^9 = 1. & & & &
\end{array}
$$

- $\mathrm{codim}_{\mathfrak{g}}[\mathfrak{g},\mathfrak{g}] = 0.$

- $\dim Der(\mathfrak{g}) = 10.$

- $\dim O(\mathfrak{g}) = 71.$

- Rank $A(\mathfrak{g}) = 6$

- Invariants of $L_{9,58}$:

$$I_1 = x_9,$$
$$I_2 = x_5^2 - 4x_4x_6,$$
$$I_3 = x_1x_5x_9 - x_4x_8^2 - x_6x_7^2 + x_5x_7x_8 + 2x_9(x_2x_6 - x_3x_4).$$

$L_{9,59}$

- Levi decomposition: $L_{9,59} = \mathfrak{sl}(2,\mathbb{R}) \overrightarrow{\oplus}_R 6L_1$

- Describing representation: $R = D_{\frac{5}{2}}$

- Structure tensor:

$$
\begin{aligned}
&C_{12}^2 = 2, \quad C_{13}^3 = -2, \quad C_{23}^1 = 1, \quad C_{14}^4 = 5, \quad C_{15}^5 = 3, \quad C_{16}^6 = 1, \\
&C_{17}^7 = -1, \quad C_{18}^8 = -3, \quad C_{19}^9 = -5, \quad C_{25}^4 = 5, \quad C_{26}^5 = 4, \quad C_{27}^6 = 3, \\
&C_{28}^7 = 2, \quad C_{29}^8 = 1, \quad C_{34}^5 = 1, \quad C_{35}^6 = 2, \quad C_{36}^7 = 3, \quad C_{37}^8 = 4. \\
&C_{38}^9 = 5.
\end{aligned}
$$

- $\operatorname{codim}_{\mathfrak{g}}[\mathfrak{g},\mathfrak{g}] = 0$.

- $\dim Der(\mathfrak{g}) = 10$.

- $\dim O(\mathfrak{g}) = 71$.

- Rank $A(\mathfrak{g}) = 6$

- Conditions on invariants:

$$
\frac{\partial F}{\partial x_j} = 0, \; j = 1,2,3.
$$

As commented before, in this case a direct integration of the system leads to some two hundred terms. In order to describe the invariants, we introduce the auxiliary functions

$$
\begin{aligned}
u_1 &= x_4^{-\frac{6}{5}} \left(5x_4 x_6 - 2x_5^2\right) \\
u_2 &= x_4^{-\frac{9}{5}} \left(25x_4^2 x_7 + 4x_5^4 - 15x_4 x_5 x_6\right), \\
u_3 &= x_4^{-\frac{12}{5}} \left(125x_4^3 x_8 - 3x_5^4 + 15x_4 x_5^2 x_6 - 50x_4^2 x_5 x_7\right), \\
u_4 &= x_4^{-3} \left(3125x_4^4 x_9 + 4x_5^5 - 25x_4 x_5^3 x_6 + 125x_4^2 x_5^2 x_7 - 625x_4^3 x_5 x_8\right).
\end{aligned}
$$

- Invariants of $L_{9,59}$:

In terms of the preceding functions, we have the (polynomial) invariants as

$$I_1 = u_4^2 + u_1 u_2 u_4 + 8 u_1 u_3^2 - 3 u_2^2 u_3 + 6 u_1^3 u_3 - 2 u_1^2 u_2^2$$

$$I_2 = u_4^4 - 100 u_1^3 u_2^2 u_3^2 - 135 u_2^4 u_3^2 + 400 u_1^4 u_3^3 + 720 u_1 u_2^2 u_3^3 - 640 u_1^2 u_3^4$$
$$+ 256 u_3^5 + 80 u_1^3 u_2^3 u_4 + 108 u_2^5 u_4 - 360 u_1^4 u_2 u_3 u_4 - 630 u_1 u_2^3 u_3 u_4 + 560 u_1^2 u_2 u_3^2 u_4$$
$$- 320 u_2 u_3^3 u_4 + 108 u_1^5 u_4^2 + 165 u_1^2 u_2^2 u_4^2 - 180 u_1^3 u_3 u_4^2 + 90 u_2^2 u_3 u_4^2 + 80 u_1 u_3^2 u_4^2 - 30 u_1 u_2 u_4^3$$

$$\begin{aligned}
I_3 = {}& u_4^6 + \tfrac{100}{13} u_1^6 u_2^6 + \tfrac{540}{13} u_1^3 u_2^8 + \tfrac{729}{13} u_2^{10} - \tfrac{900}{13} u_1^7 u_2^4 u_3 - \tfrac{6030}{13} u_1^4 u_2^6 u_3 \\
& - \tfrac{9720}{13} u_1 u_2^8 u_3 + \tfrac{2025}{13} u_1^8 u_2^2 u_3^2 + \tfrac{23880}{13} u_1^5 u_2^4 u_3^2 + \tfrac{49410}{13} u_1^2 u_2^6 u_3^2 \\
& - \tfrac{41460}{13} u_1^6 u_2^2 u_3^3 - \tfrac{117990}{13} u_1^3 u_2^4 u_3^3 - \tfrac{1215}{13} u_2^6 u_3^3 + \tfrac{33120}{13} u_1^7 u_3^4 \\
& + \tfrac{125760}{13} u_1^4 u_2^2 u_3^4 + \tfrac{9720}{13} u_1 u_2^4 u_3^4 - \tfrac{27264}{13} u_1^5 u_3^5 - \tfrac{18240}{13} u_1^2 u_2^2 u_3^5 \\
& + \tfrac{12800}{13} u_1^3 u_3^6 + \tfrac{11520}{13} u_2^2 u_3^6 - \tfrac{30720}{13} u_1 u_3^7 + \tfrac{540}{13} u_1^8 u_2^3 u_4 + \tfrac{3072}{13} u_3^5 u_4^2 \\
& + \tfrac{954}{13} u_1^5 u_2^5 u_4 + \tfrac{1620}{13} u_1^2 u_2^7 u_4 - \tfrac{2430}{13} u_1^9 u_2 u_3 u_4 - \tfrac{4140}{13} u_1^6 u_2^3 u_3 u_4 \\
& - \tfrac{1620}{13} u_1^7 u_2 u_3^2 u_4 + \tfrac{45540}{13} u_1^4 u_2^3 u_3^2 u_4 - \tfrac{14940}{13} u_1^3 u_2^5 u_3 u_4 - \tfrac{2160}{13} u_1^5 u_4^4 \\
& + \tfrac{50880}{13} u_1^3 u_2 u_3^4 u_4 + \tfrac{1215}{13} u_1 u_2^5 u_3^2 u_4 - \tfrac{60000}{13} u_1^5 u_2 u_3^3 u_4 - \tfrac{345}{13} u_1 u_2 u_4^5 \\
& - \tfrac{23040}{13} u_2^3 u_3^4 u_4 + \tfrac{69120}{13} u_1 u_2 u_3^5 u_4 + \tfrac{729}{13} u_1^{10} u_4^2 + \tfrac{1620}{13} u_1^7 u_2^2 u_4^2 \\
& + \tfrac{3810}{13} u_1^4 u_2^4 u_4^2 - \tfrac{15390}{13} u_1^5 u_2^2 u_3 u_4^2 + \tfrac{7695}{13} u_1^2 u_2^4 u_3 u_4^2 + \tfrac{24300}{13} u_1^6 u_3^2 u_4^2 \\
& + \tfrac{14175}{13} u_2^4 u_3^2 u_4^2 - \tfrac{34080}{13} u_1^4 u_3^3 u_4^2 - \tfrac{44880}{13} u_1 u_2^2 u_3^3 u_4^2 - \tfrac{6720}{13} u_1^2 u_3^4 u_4^2 \\
& - \tfrac{5855}{13} u_1^3 u_2^3 u_4^3 - \tfrac{2160}{13} u_2^5 u_4^3 + \tfrac{22500}{13} u_1^4 u_2 u_3 u_4^3 - \tfrac{10980}{13} u_1^3 u_2^2 u_3^2 u_4^2 \\
& + \tfrac{4950}{13} u_1 u_2^3 u_3 u_4^3 + \tfrac{11760}{13} u_1^2 u_2 u_3^2 u_4^3 - \tfrac{3840}{13} u_2 u_3^3 u_4^3 - \tfrac{24720}{13} u_1^2 u_2^3 u_3^3 u_4 \\
& - \tfrac{135}{13} u_1^2 u_2^2 u_4^4 - \tfrac{2070}{13} u_1^3 u_3 u_4^4 + \tfrac{1035}{13} u_2^2 u_3 u_4^4 + \tfrac{1080}{13} u_1 u_3^2 u_4^4 .
\end{aligned}$$

Replacing the u_i for the corresponding rational functions, we obtain that I_1 is a Casimir operator of order 4, I_2 of order 8 and I_3 of order 12.

$L_{9,60}$

- Levi decomposition: $L_{9,60} = \mathfrak{sl}(2,\mathbb{R}) \overrightarrow{\oplus}_R 6L_1$

- Describing representation: $R = D_{\frac{3}{2}} \oplus D_{\frac{1}{2}}$

- Structure tensor:

$$\begin{array}{llllll}
C_{12}^2 = 2, & C_{13}^3 = -2, & C_{23}^1 = 1, & C_{14}^4 = 3, & C_{15}^5 = 1, & C_{16}^6 = -1, \\
C_{17}^7 = -3, & C_{18}^8 = 1, & C_{19}^9 = -1, & C_{25}^4 = 3, & C_{26}^5 = 3, & C_{27}^6 = 1, \\
C_{29}^8 = 1, & C_{34}^5 = 1, & C_{35}^6 = 2, & C_{36}^7 = 3, & C_{37}^9 = 1.
\end{array}$$

- $\mathrm{codim}_{\mathfrak{g}}[\mathfrak{g},\mathfrak{g}] = 0$.

- $\dim Der(\mathfrak{g}) = 11$.

- $\dim O(\mathfrak{g}) = 70$.

- Rank $A(\mathfrak{g}) = 6$

- Conditions on invariants:

$$\frac{\partial F}{\partial x_j} = 0, \; j = 1,2,3.$$

- Invariants of $L_{9,60}$:

$$I_1 = 27x_4^2 x_7^2 - x_5^2 x_6^2 + 4\left(x_4 x_6^3 + x_5^3 x_7\right) - 18 x_4 x_5 x_6 x_7,$$
$$I_2 = 18 x_4 x_7 x_8 x_9 - 6 x_4 x_6 x_9^2 + 2 x_5^2 x_9^2 - 2 x_5 x_6 x_8 x_9 - 6 x_5 x_7 x_8^2 + 2 x_6^2 x_8^2,$$
$$I_3 = x_4 x_9^3 - x_7 x_8^3 + x_6 x_8^2 x_9 - x_5 x_8 x_9^2.$$

$\mathbf{L}_{9,61}$

- Levi decomposition: $L_{9,61} = \mathfrak{sl}(2,\mathbb{R}) \overrightarrow{\oplus}_R 6L_1$

- Describing representation: $R = 2D_1$

- Structure tensor:

$$\begin{array}{llllll}
C_{12}^2 = 2, & C_{13}^3 = -2, & C_{23}^1 = 1, & C_{14}^4 = 2, & C_{16}^6 = -2, & C_{17}^7 = 2, \\
C_{19}^9 = -2, & C_{25}^4 = 2, & C_{26}^5 = 1, & C_{28}^7 = 2, & C_{29}^8 = 1, & C_{34}^5 = 1, \\
C_{35}^6 = 2, & C_{37}^8 = 1, & C_{38}^9 = 2.
\end{array}$$

- $\mathrm{codim}_{\mathfrak{g}}[\mathfrak{g},\mathfrak{g}] = 0$.

- $\dim Der(\mathfrak{g}) = 13$.

- $\dim O(\mathfrak{g}) = 68$.

- Rank $A(\mathfrak{g}) = 6$

- Conditions on invariants:

$$\frac{\partial F}{\partial x_j} = 0, \ j = 1, 2, 3.$$

- Codimension three subalgebra:

$$L\langle X_1, ..X_6\rangle \simeq L\langle X_1, X_2, X_3, X_7, X_8, X_9\rangle \simeq L_{6,4}$$

- Invariants of $L_{9,61}$:

$$J_1 = x_4 x_6 - \frac{1}{4} x_5^2,$$
$$J_2 = x_7 x_9 - \frac{1}{4} x_8^2,$$
$$I_3 = 2x_4 x_9 - x_5 x_8 + 2x_6 x_7.$$

$L_{9,62}$

- Levi decomposition: $L_{9,62} = \mathfrak{sl}(2, \mathbb{R}) \overrightarrow{\oplus}_R A_{6,3}$

- Describing representation: $R = 2D_1$

- Structure tensor:

$$\begin{array}{llllll}
C_{12}^2 = 2, & C_{13}^3 = -2, & C_{23}^1 = 1, & C_{14}^4 = 2, & C_{16}^6 = -2, & C_{17}^7 = 2, \\
C_{19}^9 = -2, & C_{25}^4 = 2, & C_{26}^5 = 1, & C_{28}^7 = 2, & C_{29}^8 = 1, & C_{34}^5 = 1, \\
C_{35}^6 = 2, & C_{37}^8 = 1, & C_{38}^9 = 2, & C_{45}^7 = 2, & C_{46}^8 = 1, & C_{56}^9 = 2.
\end{array}$$

- $\mathrm{codim}_{\mathfrak{g}}[\mathfrak{g}, \mathfrak{g}] = 0.$

- $\dim Der(\mathfrak{g}) = 11.$

- $\dim O(\mathfrak{g}) = 70.$

- Rank $A(\mathfrak{g}) = 6$

- Invariants of $L_{9,62}$:

$$J_1 = 2x_4x_9 - x_5x_8 + 2x_6x_7,$$

$$J_2 = x_7x_9 - \frac{1}{4}x_8^2,$$

$$I_3 = -2x_1x_8 - 4x_2x_9 + 4x_3x_7 + x_5^2 - 4x_4x_6.$$

$L_{9,63}$

- Levi decomposition: $L_{9,63} = \mathfrak{sl}(2,\mathbb{R}) \overrightarrow{\oplus}_R 6L_1$

- Describing representation: $R = 3D_{\frac{1}{2}}$

- Structure tensor:

$$\begin{array}{llllll}
C_{12}^2 = 2, & C_{13}^3 = -2, & C_{23}^1 = 1, & C_{14}^4 = 1, & C_{15}^5 = -1, & C_{16}^6 = 1, \\
C_{17}^7 = -1, & C_{18}^8 = 1, & C_{19}^9 = -1, & C_{25}^4 = 1, & C_{27}^6 = 1, & C_{29}^8 = 1, \\
C_{34}^5 = 1, & C_{36}^7 = 1, & C_{38}^9 = 1.
\end{array}$$

- $\operatorname{codim}_{\mathfrak{g}}[\mathfrak{g},\mathfrak{g}] = 0$.

- $\dim Der(\mathfrak{g}) = 18$.

- $\dim O(\mathfrak{g}) = 63$.

- Rank $A(\mathfrak{g}) = 6$

- Conditions on invariants:

$$\frac{\partial F}{\partial x_j} = 0, \ j = 1,2,3.$$

- Invariants of $L_{9,63}$:

$$I_1 = x_4x_7 - x_5x_6,$$

$$I_2 = x_4x_9 - x_5x_8,$$

$$I_3 = x_6x_9 - x_7x_8.$$

Acknowledgement

The author benefited during the preparation of the paper from fruitful discussions with V. M. Boyko, to whom he expresses his gratitude.

References

[1] L. Abellanas, L. M. Alonso. A general setting for Casimir invariants, *J. Math. Phys.* **16** (1975), 1580-1584.

[2] N. H. Ahn and L. Van Hop. Le centre de l'algèbre enveloppante du produit semi-direct de l'algèbre de Heisenberg et une algèbre réductive, *C. R. A. S. Paris Sér. I* **303** (1986), 783-786.

[3] J. M. Ancochea, R. Campoamor-Stursberg, L. Garcia Vergnolle, Solvable Lie algebras with naturally graded nilradicals and their invariants, *J. Phys. A: Math. Gen.* **39** (2006), 1339-1355.

[4] J. A. de Azcárraga, A. J. Macfarlane, A. J. Mountain, J.C. Pérez Bueno, Invariant tensors for simple groups, *Nucl. Phys.* **B510** (1998), 657-687.

[5] J. A. de Azcárraga, A. J. Macfarlane, Optimally defined Racah-Casimir operators for su(n) and their eigenvalues for various classes of representations *J. Math. Phys.* **42** (2001), 419-433.

[6] J. A. de Azcárraga and J. M. Izquierdo. *Lie Groups, Lie Algebras, Cohomology and some Applications to Physics* Cambridge Univ. Press, Cambridge 1995

[7] E. G. Beltrametti and A. Blasi. On the number of Casimir operators associated with any Lie group, *Phys. Lett.* **20** (1966), 62-64.

[8] V. M. Boyko, J. Patera and R. I. Popovych. Computation of invariants of Lie algebras by means of moving frames, *J. Phys. A: Math. Gen.* **39** (2006), 5749-5762.

[9] V. M. Boyko, J. Patera and R. I. Popovych. Invariants of Lie algebras with fixed structure of nilradicals, *J. Phys. A: Math. Theor.* **40** (2007), 113-130.

[10] R. Campoamor-Stursberg. Invariants of solvable rigid Lie algebras up to dimension 8, *J. Phys. A* **35** (2002), 6293-6306.

[11] R. Campoamor-Stursberg. Contractions of Lie algebras and generalized Casimir invariants, *Acta Physica Polonica* **B34** (2003), 3901-3920.

[12] R. Campoamor-Stursberg. Non-semisimple Lie algebras with Levi factor $\mathfrak{so}(3)$, $\mathfrak{sl}(2,\mathbb{R})$ and their invariants, *J. Phys. A: Math. Gen.* **36** (2003), 1357-1370.

[13] R. Campoamor-Stursberg. An extension based determinantal method to compute Casimir operators of Lie algebras, *Phys. Lett. A* **312** (2003), 211-219.

[14] R. Campoamor-Stursberg. The structure of the invariants of perfect Lie algebras, *J. Phys. A: Math. Gen.* **36** (2003), 6709-6723.

[15] R. Campoamor-Stursberg. An alternative interpretation of the Beltrametti-Blasi formula by means of differential forms, *Phys. Lett. A* **327** (2004), 138-145.

[16] R. Campoamor-Stursberg. Explizite Formeln für die Casimiroperatoren des semidirekten Produktes einer Heisenberg Lie-Algebra $\mathfrak{h}$ mit der einfachen Lie-Algebra $\mathfrak{sl}(2,\mathbb{C})$, *Acta Phys. Polonica B* **35** (2004), 2059-2069.

[17] R. Campoamor-Stursberg. Affine Lie algebras with non-compact rank one Levi subalgebra and their invariants, *Acta Physica Polonica B* **38** (2007), 3-20.

[18] H. Casimir. Über die Konstruktion einer zu den irreduziblen Darstellungen halbeinfacher kontinuierlicher Gruppen gehörigen Differentialgleichung, *Proc. Roy. Acad. Amsterdam* **34** (1931), 844-846.

[19] C. Chevalley. *The theory of Lie groups.* Princeton Univ. Press, Princeton 1951.

[20] A. P. Demichev, N. F. Nelipa and M. Chai'chian. Invariant operators of inhomogeneous groups, *Comm. Math. Phys.* **90** (1983), 353-372.

[21] L. E. Dickson. Differential equations from the group standpoint, *Ann. Math.* **25** (1924), 287378.

[22] F. J. Herranz, M. Santander. Casimir invariants for the complete family of quasisimple Lie algebras, *J. Phys. A: Math. Gen.* **30** (1997), 5411-5426.

[23] A. T. Fomenko, V. V. Trofimov. *Integrable systems on Lie algebras and symmetric spaces*, Gordon and Breach, New York 1989.

[24] G. M. Mubarakzyanov. Classification of six dimensional solvable Lie algebras which possess a nilpotent basis element, *Izv. Vyssh. Ucheb. Zaved. Mat.* **35** (1963), 104-116.

[25] A. Nijenhuis and R. W. Richardson. Cohomology and deformations in graded Lie algebras, *Bull. Amer. Math. Soc.* **72** (1966), 1-29.

[26] J. C. Ndogmo. Properties of invariants of solvable Lie algebras, *Canad. Math. Bull.* **43** (2000), 459-471.

[27] J. C. Ndogmo. Invariants of a semi-direct sum of Lie algebras J C Ndogmo *J. Phys. A: Math. Gen.* **37** (2004), 5635-5647.

[28] P. J. Olver. *Classical Invariant Theory,* London Math. Soc., London 1999.

[29] J. Patera, R. T. Sharp, P. Winternitz and H. Zassenhaus. Invariants of real low dimension Lie algebras, *J. Mathematical Phys.* **17** (1976), 986-994.

[30] J. Patera, P. Winternitz and H. Zassenhaus. Quantum numbers for particles in the De Sitter space, *J. Math. Phys.* **17** (1976), 717-728.

[31] J. Pecina-Cruz. An algorithm to calculate the invariants of any Lie algebra, *J. Math. Phys.* **35** (1994), 3146-3162.

[32] J. Pecina-Cruz. On the Casimir of the group ISL(n,R) and its algebraic decomposition, *J. Math. Phys.* **46**, N.6., 2005, p.063503.

[33] A. Peccia, R. T. Sharp. Number of independent missing label operators, *J. Math. Phys.* **17** (1976), 1313-1314.

[34] A. M. Perelomov, V. S. Popov. Casimir operators for classical groups, *Dokl. Akad Nauk SSSR* **174** (1967), 287-290.

[35] V. S. Popov, A. M. Perelomov. Generating functions for Casimir operators, *Dokl. Akad. Nauk. SSSR* **174** (1967), 712-714.

[36] C. Quesne. Casimir operators of semidirect sums of Lie algebras, *J. Phys. A: Math. Gen.* **21** (1988), L321-324.

[37] G. Racah. Sulla caratterizzazione delle rappresentazioni irriducibili dei gruppi semisemplici di Lie. *Atti Accad. Naz. Lincei. Rend. Cl. Sci. Fis. Mat. Nat.* **8** (1950), 108-112.

[38] G. Racah. *Group Theory and Spectroscopy*, New Jersey: Princeton Univ. Press, 1951.

[39] M. Rhomdani. Classification of real and complex nilpotent Lie algebras of dimension 7, *Linear and Multilinear Algebra* **24** (1989), 167-189.

[40] H. Stephani. Differentialgleichungen. Symmetrien und Lösungsmethoden. *Spektrum Akad.* Verlag, Berlin 1994.

[41] S. Tremblay, P. Winternitz. Invariants of the nilpotent and solvable triangular Lie algebras, *J. Phys. A: Math.* Gen. **34** (2001), 9085-9099.

[42] V. V. Trofimov. Semi-invariants of a coadjoint representation of Borel subalgebras of simple Lie algebras, *Trudy Sem. Vektor Tenzor. Anal.* **21** (1983), 84-105.

[43] V. V. Trofimov. Finite dimensional representations of Lie algebras and completely integrable systems, *Mat. Sb.* **3** (1980), 610-621.

[44] P. Turkowski. Low-dimensional real Lie algebras, *J. Math. Phys.* **29** (1988), 2139-2144.

[45] P. Turkowski. Structure of real Lie algebras, *Linear Algebra Appl.* **171** (1992), 197-212.

[46] L. Šnobl, P. Winternitz. A class of solvable Lie algebras and their Casimir invariants, *J. Phys. A: Math. Gen.* **38** (2005), 2687-2700.

[47] E. Weimar-Woods. Contractions, generalized Inönü-Wigner contractions and deformations of finite dimensional Lie algebras, *Rev. Math. Phys.* **12** (2000), 1505-1525.

[48] H. Zassenhaus. On the invariants of a Lie group, in *Computers in Nonassociateive Rings and Algebras (Special Session, 82nd Annual Meeting AMS 1976)*, Editors R. E. bock and B. Kolman, NY, Academic Press 1977, 139-155.

Index